SOLDIER REPATRIATION

Soldier Repatriation
Popular and Political Responses

KAARE DAHL MARTINSEN
The Norwegian Institute for Defence Studies, Norway

Routledge
Taylor & Francis Group
LONDON AND NEW YORK

First published 2013 by Ashgate Publishing

Published 2016 by Routledge
2 Park Square, Milton Park, Abingdon, Oxfordshire OX14 4RN
711 Third Avenue, New York, NY 10017, USA

First issued in paperback 2016

Routledge is an imprint of the Taylor & Francis Group, an informa business

British Library Cataloguing in Publication Data
A catalogue record for this book is available from the British Library

The Library of Congress has cataloged the printed edition as follows:
Dahl Martinsen, Kaare.
Soldier repatriation : popular and political responses / by Kaare Dahl Martinsen.
pages cm
Includes bibliographical references and index.
ISBN 978-1-4724-1649-0 (hardback)
1. Afghan War, 2001---Repatriation of war dead--Great Britain. 2. Afghan War, 2001---Repatriation of war dead--Denmark. 3. Afghan War, 2001---Repatriation of war dead--Germany. 4. Afghan War, 2001---Casualties--Great Britain. 5. Afghan War, 2001---Casualties--Denmark. 6. Afghan War, 2001---Casualties--Germany. 7. Afghan War, 2001---Press coverage--Great Britain. 8. Afghan War, 2001---Press coverage--Denmark. 9. Afghan War, 2001---Press coverage--Germany. 10. Afghan War, 2001--Public opinion. I. Title.
DS371.412.D34 2013
958.104'76--dc23

2013020855

ISBN 13: 978-1-138-27935-3 (pbk)
ISBN 13: 978-1-4724-1649-0 (hbk)

Contents

Preface

The idea for this book came from watching a repatriation ceremony on television with my students. Most of them had been to Afghanistan, and their comments revealed how deeply affected they were by what they saw. What puzzled me was their attention to details. Without their help, certain tell-tale aspects of the ceremony would have passed me by, and this book would never have been written.

In my work, I have been able to draw on the insights of students from many countries, including the three singled out here. Writing has been an arduous and at times emotionally challenging process. Without the close reading, criticism and encouraging comments of my colleague Kjell Inge Bjerga, it would have taken longer.

Kaare Dahl Martinsen
The Norwegian Defence University College

List of Abbreviations

Bmvg	Bundesministerium der Verteidigung [Federal Ministry of Defence]
CDU	Christlich Demokratische Union [Christian Democratic Union] The main German conservative party
CSU	Christlich Soziale Union [The Bavarian sister party of the CDU]
Dpa	Deutsche Presse-Agentur [German news agency]
FAZ	*Frankfurter Allgemeine Zeitung*
IED	Improvised Explosive Device, roadside bomb
ISAF	International Security Assistance Force
MoD	Ministry of Defence
RAFPA	Royal Air Force Police Association
RoE	Rules of Engagement
SPD	Sozialdemokratische Partei Deutschlands [Social Democratic Party of Germany]

Translations

Unless otherwise noted, all translations in the text are my own.

Chapter 1
Private Grief and Public Mourning

The number of coalition soldiers sent back in coffins from Afghanistan has increased steadily since the war began in 2001. From 2005 the rise was particularly steep, pushing the total number of dead beyond 2,400 by the end of 2011 (iCasualties 2013). The trend has not levelled out with the approach of the scheduled exit in 2014. Nevertheless, public interest in the death toll has dropped. The first casualties in Afghanistan were front-page material.[1] By 2008 the deaths had become less newsworthy. Only a large number of deaths or particularly gruesome details now ensure full media coverage. In March 2012, when six British soldiers travelling in an armoured vehicle were killed instantly by an IED, it made the front pages of British newspapers. The death of an airman 12 days earlier was barely noticed (Gerges 2012).[2]

The extensive media coverage of troop deaths in the early years derived partly from the surprise felt by participating nations. They were unprepared for a rising casualty count. In 2001, it was widely expected that the Western deployment would be of short duration and that losses would be few. The Taliban regime was portrayed as singularly oppressive. Western soldiers in concert with domestic groups in Afghanistan would oust the regime and pave the way for a transition to a more democratic rule. A new leadership would be installed that was strong enough to prevent the return of terrorists seeking to use the country as a training ground. Toppling the Taliban regime would not take long; the fall of Kabul went faster than imagined, and this was taken as proof the troops would not have to stay long. Tribal leaders eager to co-operate with the West during initial military operations gave the impression of an alternative Afghan elite ready to assume power. As soon as a new government was installed, the Western coalition could withdraw. 'If I had known then that a decade later we would still be fighting in Afghanistan, I would have been profoundly perturbed and alarmed,' Tony Blair wrote in his memoirs (Blair 2010: 347).

What in 2001 had spurred the West to deploy troops is easily summed up. The prime reason was to prevent terrorism from spreading and ultimately engulfing the West. Another motive very palpable at the start of the war was a sense of solidarity with the United States. For the first time in its history, NATO invoked Article 5 stating that '9/11' had been an attack not just on the United States but on all members. Even in countries where support for the US had never been particularly strong,

1 Casualties will be used here when referring to deaths. In military parlance, casualties cover both the dead and the wounded.

2 By comparison, neither *The Times* nor *The Guardian* mentioned his death.

expressions of transatlantic unity ran high, perhaps best expressed on *Le Monde's* front page two days after the twin towers collapsed: '*Nous sommes tous Américains*' [We are all Americans] (2001). Secondly, there was a widespread and strong sense of sympathy for the United States. Chancellor Gerhard Schröder expressed 'unlimited solidarity' with the Americans (*Welt* 2001). In Denmark, Prime Minister Poul Nyrup Rasmussen declared that this was 'an attack on the entire democratic world' (Rasmussen 2001). Tony Blair echoed the same sentiments in a speech given just after the start of military operations: 'We know the al-Qaeda network threatens Europe, including Britain, and indeed any nation throughout the world that does not share their fanatical views' (Blair 2001). The media in all three countries interpreted the attacks on the United States as directed against the West and not just the Americans, and that new threats were imminent (Hammer 2008, Bromley and Cushion 2011). The willingness to react was strong. Since the end of the Cold War, deployments abroad had grown both in time spent and personnel sent. Afghanistan was distant, but no farther away than the Democratic Republic of Congo, where the Europeans had undertaken a peacekeeping mission a few years before. Life at home would not be perturbed; in past military operations involving the Europeans, loss of life had been minimal. The heaviest burden would be economic. Extra funding was made available in defence budgets, and sums were earmarked for reconstruction and civilian assistance.

Soon, developments on the ground in Afghanistan frustrated Western expectations. Catching the Taliban leadership was difficult and the heads of Al Qaeda managed to evade capture for years. The porous border with Pakistan and the continued recruitment of young men to fight against the Western forces became a major headache. The worsened security situation made reconstruction difficult, adding to the problems caused by a corrupt Afghan government. From the middle of the decade it became clear that the war was not drawing to an end, prospects of success were dim, and casualties were rising. This led to a downscaling of ambition for the Western presence in Afghanistan. Talk of a democratic regime with an accountable and transparent system of governance and a growing economy ceased. Instead, the aim was reduced to ensuring the transfer of security control from Western forces to Afghan units. Although this has meant a gradual reduction in combat operations, the number of dead bodies sent back has not diminished.

This book is about how the soldiers killed in Afghanistan have been repatriated, received and commemorated in three European countries: Britain, Denmark and Germany. There are no national differences in how family and friends experience the loss of a soldier. When a soldier is shot to death or killed by a roadside bomb in Afghanistan, the reactions of fellow unit members near the scene and of family and friends upon receiving the death message tend to fall within the same range, irrespective of nationality. It is only when the death becomes a public event that differences emerge. When returning coffins were transported through the village of Royal Wootton Bassett in the south of England, the inhabitants honoured the dead by standing quietly on the pavement as the coffins passed. In Germany, paying tribute in a similar way would have been practically impossible. Not only were the routes of the coffins generally unpublicised, but ceremonies were conducted in military encampments closed to the public or only in the presence of family. At the political level, paying homage to the dead was long avoided in Germany, most likely to prevent accusations that the country had reverted to militaristic hero worship. Only gradually did the practice become more open. In Denmark, official receptions from the start were more publicly ceremonial than in the two other countries. This may seem surprising, given the Danes' lack of military tradition. But for Denmark, participation in the war was part of a larger security policy transformation the country had initiated in the early 1990s. One key aspect of it was the active use of Danish armed forces abroad to gain a stronger international role. While hero worship was shunned in Germany, it was actively encouraged in Denmark.

The choice of countries to be examined in this book requires some explanation. Common to all three was an ambition to use deployment in Afghanistan as a way to boost political influence. Britain's ambition was to prove the importance of its transatlantic partnership with the United States. Denmark was similarly motivated. Its foreign deployments would become a prime bargaining chip to improve contacts with the United States and Britain, while Germany intended its participation as a visible expression of German reliability as a security partner. But there are few other similarities. The number of casualties suffered by each country in Afghanistan reflects the different roles they assumed and the limits each government felt it must impose on troop commitments to avoid a political backlash. Casualties are more revealing than troop deployment sizes. Britain and Germany are the second- and third-largest contributors to the war in terms of personnel, with 9,500 and 4,900 soldiers respectively (ISAF 2012). As for losses, by March 2013 Britain's death toll of 440 was second to the 2,178 lives lost by the United States (iCasualties 2013). Denmark's 750-strong contingent is the largest sent by any country in relation to the size of its armed forces. Forty-three Danish soldiers have been killed; no other coalition member has lost a higher percentage of its deployed forces. German has lost 53 soldiers, a relatively small tally compared with the number of troops deployed and the size of the German population. However, such simple comparisons provide little insight into the war's impact on political debate inside the three countries. In Germany the government had a hard time gaining acceptance for the deployment in the Bundestag, or parliament, and in the population at large. As a consequence, when dead bodies started coming home the political and military leadership's response was to remain as low-key as possible. In Denmark, political consensus in favour of the mission meant minimal public debate, a state of affairs the government encouraged by treating the dead as symbols of a regained national pride.

Death and Strategic Culture

Afghanistan is a reminder that wars entail losses. For each of the three countries studied, no other conflict of the past two decades has inflicted nearly as many casualties as Afghanistan. The 1990–91 Gulf War brought surprisingly few losses of coalition troops. During Operation Desert Storm, after six weeks of fighting, the 795,000-strong coalition force had only lost 240 men (Biddle 2004). Similarly, the Libyan operation conducted in 2011 resulted in zero casualties.[3] The impression became entrenched that the West could fight wars without losses. The reverence of the media for high tech weapons like precision bombs reinforced the idea of Western invulnerability and superiority. The shift in perception of war's meaning was identified with considerable foresight by Don DeLillo in *End Zone*. In this novel, the main character observes: 'War has always told men what they were capable of under stress. Now it informs the machines. It's the best test of a country's technological skills. ... War brings out the best in technology' (DeLillo 1975: 84–5). Afghanistan has shattered the impression that wars can be fought without casualties. The dead have brought the war home. The remoteness of war from the attention of ordinary people is well captured by the titles of the French philosopher Jean Baudrillard's essays about the Gulf War: 'The Gulf war will not take place', 'The Gulf War is not really taking place', and finally 'The Gulf War did not take place' (Baudrillard and Patton 1995). His argument was that the fighting taking place was so far removed from what most people imagined as combat that it was difficult to label it war at all. The fact that the number of US motorists killed in traffic accidents exceeded the number of US soldiers killed in the war during the same time span meant that the Western way of warfare was an atrocious masquerade. In a way it had become victimless and thus '... does not seem like a real war but rather a prefiguration of an experimental, blank war, or a war even more inhuman because it is without human losses'. But Afghanistan has turned out differently, belying the philosopher's claim that the West's use of sophisticated weaponry had deprived war of its contents. Afghanistan is in many ways a traditional war, something most Europeans thought obsolete.

Security scholars, especially European ones, were among those caught unprepared. Numerous books and articles have been written about the perceived growth of a common European strategic culture, outlining how the different countries' perceptions and reactions to security threats display more and more commonality (Sondhaus 2006). A search for comments about the role of mutilation and death in this process turns up few results. The explanation is not difficult: none of the European Union military operations have involved prolonged combat, and none have resulted in anything comparable to the steady stream of body bags flown home from Afghanistan. Moreover, no EU operations have triggered wide-scale debate, political challenge or government self-justification at home. Convergence towards a common strategic culture is usually measured by comparing political statements and communiqués made in the run-up to a crisis (Meyer 2006, Gross 2009). In short, the EU has formulated an increasingly ambitious security role for itself on paper, but so far has very little to show for it (Matlary 2009, Merchet 2009, Toje 2010). The apparent unity makes it easy to overlook national differences. In a

3 Strictly speaking, the US airman killed in a traffic accident in Italy may be counted as the operation's only casualty.

war, these differences emerge when dead soldiers are brought back home and the deaths are recognised publicly and given a political justification.

Exploring the national response to casualties will therefore add to our understanding not only of national security policies but also of alliance cohesion, or more precisely the lack of it. Traditionally, when cohesion has been discussed within NATO or analysed by outside scholars, the focus has been on burden sharing, insufficient modernisation of equipment, and the shortage of units that can be deployed at short notice. All these problems are well known. Still undocumented is the impact of war deaths on a country's willingness to contribute to military operations riskier than peacekeeping. Although the governments have voiced broadly similar official justifications when the coffins are received back home, the domestic debates triggered by combat deaths have varied.

One might have assumed such differences to be due to the extent of political opposition at home. But this was not the case, with governments and the main opposition parties in all three countries agreeing to the deployment, reinforcement and maintenance of troops in Afghanistan. In fact, despite the growing number of dead, Afghanistan has neither been a constant nor frequent concern in any of the three countries' parliaments. To gain a fuller picture of the national differences concerning casualties and strategic culture, it is necessary to go wider and include debates conducted in the media. In addition to interviews and articles featuring politicians and security experts, a third group of commentators has emerged: the men and women coming back from Afghanistan. Their views carry weight, and they have had the ability to influence not only politicians but, more importantly, public opinion.

Although public opinion is an enduring concept when casualty sensitivity is analysed, how to define it, let alone measure it, remains notoriously difficult (Smith 2005). In 1961, V.O. Key wryly concluded: 'To speak with precision of public opinion is a task not unlike coming to grips with the Holy Ghost' (1961: 8). Nothing as challenging will be attempted here. Opinion polls will be used as an indicator of popular attitudes towards the war. This is not unproblematic. Sometimes they have been commissioned by newspapers whose editorial line strongly influences the wording of the questions or the interpretation of the results. Especially at the beginning of the war, the pro-war atmosphere led to distorted interpretations. When the Scottish tabloid *Sunday Mail* published the results of its own poll showing only 5 per cent favouring bombing and 69 per cent preferring some form of peaceful conflict resolution, the headline was 'Scots split on bombing' (Lavery 2001). In this book, opinion polls will be referred to but will not play a key part. First of all, the questions posed have varied, making comparisons over time and across borders difficult. The one most frequently asked – 'Are you in favour of withdrawing British/Danish/German troops from Afghanistan?' – yields very different answers if people are given alternatives. When polled in July 2009, the strong majority opting for a return of British soldiers was split when asked if the withdrawal should be immediate or dependent upon stability in Afghanistan (BBC 2012). The portion going for the latter alternative could be taken as supportive of the government line.

Nonetheless, negative polls and bad news often triggers a response from politicians and military leaders trying to influence how the public should respond. When the first 100 British soldiers had been killed in Afghanistan, the Minister of Defence warned against letting such numbers 'deflect us' from the importance of the war (Ainsworth 2009). Virtually the same was said by his successor three years later, when the total number of casualties had more than doubled and public support for the war had dropped to an all-time low (Hammond 2012).

What the ministers attempted was to influence public opinion, raising the question of whether public support is event-driven or elite-driven: is it the sudden rise in casualties or the official response to the bloodshed that influences how people respond? Clearly, the two interplay. When soldiers are killed by a roadside bomb or a Western hostage is decapitated, no elite signals are required to influence popular opinion. That does not mean political leaders are without sway. Douglas Kriner and Graham Wilson (2010), two American political scientists, have shown that when survey participants were told in advance that all political leaders agreed on remaining in Afghanistan, the responses aligned with elite opinion. Politicians are well aware of that effect and have tried hard to maintain and boost support for the war through their statements, especially at times of rising casualties. Casualties influence public opinion, and public opinion influences any government's ability to conduct war. The Spanish withdrawal from the Iraq War in 2004 and the Dutch from Afghanistan in 2010 show the power of public pressure. Likewise, François Hollande made the immediate withdrawal of France's contingent in Afghanistan a key point of his successful 2012 campaign for the presidency. This is what casualty sensitivity refers to. It should be underlined, though, that it's a neutral concept. Casualty sensitivity also exists in countries that have maintained their presence in Afghanistan; the question posed in this book is how robustly national politicians counter it in defence of the mission.

In Germany, casual sensitivity is sometimes equated with 'casualty shyness' (Kümmel and Leonhard 2004). This is an apt description of the response displayed by many governments, Germany's included, when the number of dead rises. Fearing that casualties will undermine public opinion, political leaders may try to soften the effect of combat deaths in the manner of the British minister quoted above or even to minimise casualties for the express purpose of preserving support by changing the national force's Rules of Engagement (RoE). An example of the latter emerged in early 2011 when the German government announced it would radically redraw the Bundeswehr's strategy in Afghanistan, aiming to shrink the area controlled by German units (*Die Welt* 2011). Behind the change was the assumption that a reduced presence would diminish exposure to Taliban attacks. An increasing casualty rate, it was felt, could undermine public support to such a degree that withdrawal would be the only way for the government to retain its hold on power.

There is nothing new in such behaviour. The American security scholar J. Lee Ray (1995) has summed it up thus: '[P]ublic opinion in democracies will exert a pacifying impact, in part because the masses pay the heaviest cost of interstate wars.' This is not very different from Immanuel Kant's reasoning (2003: 9), more than two centuries earlier, that if allowed to decide whether to unleash a war or not, citizens would '... consider all its calamities, before committing themselves to so risky a game'. The problem is that nobody can predict the costs in advance; the quotation from Tony Blair's memoirs shows that this applies even to the best informed.

The Body-Bag Syndrome

While the European lack of attention to the impact of casualties on security policies has been mentioned, the same lacuna is evident in connection with the relationship between casualties and public opinion. That most of the experts quoted so far are American is therefore no coincidence. American researchers have looked deeply into the impact of dead soldiers on popular support for military action and the effects on political decision-making. Much of the discussion below builds upon the conclusions drawn by American researchers.

The American historian and military expert Edward N. Luttwak (1996) has concluded that the probable magnitude of US casualties will always top the agenda when Pentagon officials and military officers discuss a possible intervention by the United States. Their fear is that the public will demand immediate withdrawal at the first sight of body bags coming home. US politicians and generals are not the only ones to have made that connection. Shortly before the UN intervention in Somalia in 1992, a Somali leader told US special envoy Robert Oakley: 'We have studied Vietnam and Lebanon and know how to get rid of the Americans, by killing them so that public opinion will put an end to things' (Reiter and Stam 2002). He was right; despite US numerical strength and high-tech equipment, the dragging of dead US soldiers in the dust before running cameras had an immediate impact on the public back home. Likewise, during the Kosovo War, the Serb threat to fight any ground invasion with maximum force coerced the US to go for air strikes instead of ground troops (Posen 2000). When the Dutch government withdrew from Afghanistan in 2010, a Taliban leader sent out a congratulatory press release and warned at the same time that a possible replacement of the military contingent with development aid workers would not lessen the Taliban's efforts to annihilate them (Chin-A-Fo 2010).[4]

The withdrawals would seem to support the basic assumption in casualty sensitivity theories that casualties undermine public support for war. Before pursuing that line of thought in greater detail, the opposite phenomenon should be considered: that combat deaths can have a rallying effect on public opinion. Victory gives meaning to the sacrifice; as World War II in the Pacific neared an

4 See also the remarks made by the British Foreign Secretary's Special Envoy to Pakistan and Afghanistan on the Taliban's reading of Western election politics (Foreign Affairs Committee 2011: 169).

end, US casualties increased but so did public support (Mueller 1973). Several politicians have argued along these lines, and indeed, periods of rising support after a surge in deaths have been observed in Afghanistan (Norton-Taylor et al. 2009). These developments have been temporary, however. In all three countries, support for the war is lower than it was at the start. An important reason is that after more than a decade, a decisive victory is not in sight.

Research has shown that successful military operations enjoy public support even in the face of high casualties (Jentleson and Britton 1998). The US experts on civil–military relations Peter Feaver and Christopher Gelpi, analysing data covering the period from the Korean War to the operations in Kosovo in 1999, concluded that if victory seemed likely, decision makers could rely on substantial popular support despite a high death toll (2004: 70–46). The same conclusion has been drawn by the American political scientist Richard Eichenberg: 'The public is defeat phobic, not casualty phobic' (2005: 140–77). On this crucial point the war in Afghanistan has been a challenge for decision makers. Ministers defending the war have often cited the significant progress achieved in education, especially for girls, and in health care in areas controlled by Western forces. Nevertheless, such achievements pale in the media against suicide attacks and roadside bombings. With no clear criteria for the measurement of success, the number of dead has played a defining role in forming public opinion (Johnson and Tierney 2006). Peace researchers have long bemoaned what they see as a tendency by the media to overemphasise bad news, while positive developments go unreported simply because they are dull (Boulding 1978, Mack 2005). In Afghanistan, acid attacks on girls walking to school would get more attention than the opening of their school did. The lack of media attention to progress in Afghanistan has been a continuing concern to politicians and military leaders. School construction and well drilling cannot compete for news coverage against a rising death toll, simply because casualties have stronger emotional impact (Newman and Perloff 2004). The emotive power of casualty reports makes them accessible to a wider public than other news reports. Death in the field is sudden and random. Moreover, soldiers tend to share traits with the populations they represent, making articles about the dangers they face that much more compelling. No previous knowledge or insight is required; readers do not have to know why their country is fighting in Afghanistan or to hold any particular policy opinions. Casualties create a large, attentive public in a way that no other reports coming out of the war can. Debates about reconstruction, corruption and opium production have select audiences whose interests may be as narrow as a single issue. Casualties affect everybody.

The duration of a war, like the degree of success achieved, has significant impact on popular attitudes. As interest in Afghan events dropped off in tandem with support for the war, a Danish journalist concluded in 2010 that the population was suffering from 'war fatigue' (*Jyllands-Posten* 2010a). He would have found support for this conclusion in the writings of John Mueller, an American political scientist who was among the first to analyse the relationship between casualties and public opinion. The longer a war lasted, Mueller (1971) concluded, the more likely support would tend to fall. In both the Vietnam and the Korean wars, support dropped as casualties grew, especially when the number took a sudden jump. Yet in the case of the Vietnam War, support fell at a far slower pace than casualties grew, suggesting that other factors were at play. Mueller (1973) provides one explanation: that the public is far more sensitive to losses at the beginning of a war, and will react only to large ones later on. If that were so, the steep increase in losses from 2005 should have had a distinct impact on popular backing. None of the polls taken at the time bear this out.

It should be underlined that Mueller drew his conclusions from conventional wars; other kinds of armed conflicts were overlooked. This is a major limitation given that conventional wars are not only rare but increasingly unlikely to break out. Unless the scope is widened to include less spectacular uses of military force, it remains impossible to explain why some wars or military actions have garnered less support than others. For instance, the US decision to intervene in Korea in 1950 was met with greater understanding from the public than sending troops to Somalia in 1992; a communist Korea was perceived as a threat while the importance of Somalia to US security was less apparent. Thus, if military action is seen as vital to the defence of national security, support tends to grow. The problem is that opinions differ widely on how national security is best defended (Schörnig 2007, Martinsen 2010). When German Defence Minister Peter Struck (2004) claimed in parliament that Germany was being defended in the foothills of the Hindu Kush, many of his countrymen responded with bafflement and ridicule. A few years earlier, the German perspective had been different. Foreign Minister Joschka Fischer (1999) of the strongly pacifist Green Party managed to gain support for joining NATO's attack on Serbia by arguing that it was in the country's national interest to prevent a repetition of the war crimes perpetrated half a century earlier by the Wehrmacht. The difference between the two cases is national interest. Fischer managed to make that link plain to most; Struck was less successful.

What Fischer did was to use national interest as an argument in favour of war even though there was no threat to German national security. This line of arguing is often used when countries agree to allocate soldiers for peacekeeping missions. Even when such a decision enjoys popular backing, the degree of support will usually drop if peacekeepers are killed.[5] One reason may be the gap between the popular perception of peacekeeping – namely, that it is peaceful – and the more troubling reality on the ground. The way out for a government may be to reclassify the operation from peacekeeping to peace enforcement, stabilisation or aid to the civilian population. A striking example was the German government's insistence that the Afghanistan deployment was primarily one of assistance until finally it admitted that the Bundeswehr was fighting a war.

5 Steven Kull (1995) has claimed that casualty sensitivity has been exaggerated for peace-keeping operations. However, the data he provides do not indicate that public tolerance approximates the levels seen when national security or interests are at stake.

The perception the German government wished to convey was eventually refuted by the number of dead being sent back home. Another problem was a disconnect between the official benign image of the country's armed forces and violent events on the ground. Such a gap can have a drastic effect on public support. In September 2006, photos of German soldiers posing with a skull and bones in Afghanistan hit the front page of most German newspapers. Although the armed forces argued on little evidence that the skull was likely Russian (presumably dating from the Soviet occupation of the 1980s) the reputation of the Bundeswehr suffered a blow. Claims were made that Germany now had its own Abu Ghraib scandal, and the impact on public opinion was noticeable (*Bild* 2006). The share of the populace reporting a negative impression of the Bundeswehr almost doubled in the course of a year, from 17 per cent to 32 per cent (Bulmahn 2008). Support for German participation in the International Security Assistance Force (ISAF) in Afghanistan dropped to 40 per cent from 64 per cent the preceding year (Fiebig and Pietsch 2010).

The opposite effect, of climbing support, may occur if national troops win a battle or manage to kill a Taliban commander. The Danish defence of the village Musa Qala against repeated Taliban attacks over a 36-day period in July and August 2006 is a good example. The soldiers were portrayed as heroes by the media and politicians alike, and public support for Danish participation rose. The effect was amplified by depictions of the battle as one in which the smaller, more vulnerable party won despite being outnumbered. In Danish history, the opposite had always been the case (Ringsmose 2007).

A key to popular support and thus casualty acceptance is the political elite. A signal of determination and unity from the elite has great persuasive power on a population. In a case study of the Vietnam War, the American media researcher John R. Zaller showed that only after the US elite's consensus cracked did the public become polarised (1992). Building on this, his fellow countryman Adam J. Berinsky (2007) showed that disagreement among elites need not be widespread to create public division; a few dissenting voices can be enough to fortify the opposition. When the British Foreign Secretary Robin Cook quit his post over British engagement in the Iraq War in March 2003, only to be followed a few months later by the resignation of Clare Short, the Secretary of State for International Development, and still later by the protestations of former cabinet member Mo Mowlam, the discord seriously damaged Tony Blair's effort to portray the Labour Party leadership as united.

Infighting at the top immediately conveys to those outside that the necessity of war is disputed. People opposing it discover they are not alone and will find it easier to express their opinions (Noelle-Neumann 1974). The media will be able to substantiate critical articles with reference to people who belong to the elite, and therefore should be in the know. There will always be politicians, academics, and artists opposing the government line, but their voices are feeble compared with the impact of a political leadership in disarray. That is why the prime ministers in all three countries examined here have emphasised, without exception, that the decision to send troops to Afghanistan rested on a strong political majority. Rarely mentioned is the fact that it took considerable arm-twisting and the threat of a parliamentary confidence vote to obtain majority support in Germany, while in Britain the parliament was merely informed and not asked to vote. Dissenting voices have occasionally been heard from the largest opposition parties and interest groups, but even in these cases the response from the political leadership has been strong. The importance of maintaining unity has been a priority for all the governments. What will be shown in Chapter 3 is that even when unity at the top helps ward off criticism of the decision to go to war, criticism of the war's conduct can still flare up with damning effect on a government.

Framing Death

The challenge facing politicians before and during a military campaign is 'to persuade one's own people, and those of one's allies that the benefits of military action, or preparation for it, will outweigh the costs' (Howard 1982/3). The different aspects of casualty sensitivity referred to above are all based on the premise that, from the day a war starts, the population balances the costs and benefits, then shifts its support based on 'whether the outcome is judged worth the costs – all informed by leaders and experts' (Larson 1996). The catch is that very few people have an adequate basis to assess costs, especially when the economic burden and loss of life are quantified in uncertain future terms (Gilens 2001). The uncertainty does not keep people from weighing costs and benefits as they make up their mind; what merits closer attention is how opinion makers frame the information the public relies on.

In a news context, a frame is a journalistic device that brings certain elements of political reality into focus over others. This means that some aspects or details are blurred or left out while others are emphasised if not exaggerated. Clearly, journalists are not alone in using frames; politicians and interest groups also employ framing to get their message through (Lawrence 2000). Framing refers to how something is presented. The aspects that are accentuated influence how readers and viewers perceive a problem and judge how it should be solved (Entman 1993). The number of communication advisors employed by politicians and ministries gives an indication of how important framing is to the political process. The defence sector has also been part of this trend, creating web pages and blogs to present its view of important issues, including the war in Afghanistan. Facts supporting the official line are presented. For instance, German, British and Danish official blogs all contained articles in May 2012 outlining the successful transfer of security control to Afghan forces (Forsvarsministeriet 2012, Laing 2012, Meyer 2012). Their similarity was evident from the almost identical photos of smiling Afghan soldiers in clean uniforms accompanying the articles. Despite the growing number of security experts, ex-soldiers and journalists in all three countries who questioned the capability of the Afghans and the pace of coalition progress, articles published on the ministry websites bore no suggestion of doubt.

Facts or views contradicting the frame are a threat to the version of reality political leaders wish to convey. Two examples show how. One concerns the official German line portraying soldiers as 'Bürger in Uniform' [citizens in uniform], implying that they are no different from the rest of society and, equally important, that they are different from the Germans who served in Hitler's Wehrmacht. Thus when the photos of Bundeswehr members posing with human skulls were published, the citizen-in-uniform myth cracked and the resemblance to photos showing Wehrmacht soldiers posing with dead civilians as trophies was uncomfortably strong (FAZ 2006). The other example is not limited to Germany but common to all the three countries. Their governments framed the war as one in which the vast majority of Afghans opposed the Taliban. Only in the mid-2000s did tribal and ethnic conflicts surface as explanations for the continued fighting, especially in the south of the country. Western populations might accept casualties in a war against bin Laden and his men, but a purely internal conflict was less likely to be seen as worth the cost in blood. The willingness to risk the lives of one's own countrymen in a distant civil war is at best limited. The civil war in Yugoslavia was an exception to this rule, because potential spill-over effects on European security were obvious. In Afghanistan it was logical for officials to frame the initial Western deployment as imperative to check a growing terrorist peril that ultimately could harm us. Tribal aspects of the conflict were downplayed, if mentioned at all, in the official brochures issued by the Danish, German and British governments.

Public statements justifying the war contain explicit frames. When the Danish Prime Minister gave a speech in Afghanistan presenting the opening of a local girls' school as a victory, the photos of him surrounded by Danish soldiers in combat fatigues in Afghanistan made it clear to the reader that the school would not have been opened had the Danes not been there (Svane 2007). Less explicit framing devices include newspaper captions, the style and size of print, and the level of priority assigned to a story by television news and newspaper editors. Different forms of political behaviour can also be seen as framing devices. When the Danish Defence Minister personally attended the reception ceremony marking a dead soldier's return home, the act highlighted the importance the government attached to the war. This can be contrasted with the German government's failure over time to come up with a consistent reception procedure open to the media.

Frames do not have to be visible to be influential; unseen ones are no less important. Their presence is sensed but remains 'difficult to locate, to put your finger on' (Watson 2007: 107). One reason is that most framing is not overtly argumentative. It is not the government asking you to agree with its analysis or justification; it is the use of emotive symbols and images to trigger your emotional response. One example has been the use of photos showing the homecoming of coffins in Royal Wootton Bassett, with villagers standing silently in respect on the pavements. When ordinary people respond to the dead in this way, the persuasive power is great.

The overriding aim of decision makers has been to frame the news in a way that ideally supports the war effort, or at least does not undermine it. This does not always succeed because frames, seen or unseen, have a divisive potency. When new Bundeswehr recruits swear their oath of allegiance to the German constitution in front of the historical Reichstag building with the Chancellor and government attending, some will draw lines back to 1930 and the militarism associated with the Third Reich. Others will see the ceremony as an important coming-of-age for the new democratic Germany.

Media experts sometimes question why one frame is chosen and not another (Norris et al. 2003: 11). When it comes to political communication, that question is easily answered. The frame most likely to guarantee support for a specific policy will be the one chosen. For the authorities, it is essential that their version of events is seen as credible. According to William Merrin, a media analyst, credibility has replaced truth as the guiding principle of the news (Merrin 2005: 71). The public lacks the means of verifying the matters communicated, and will often have to base its judgement on whether the communicator is credible or not. Thus, in many cases when the official version is attacked, the official reply will contain not only a rejection but also a statement questioning the credibility of the critics. This was the case when Danish public television broadcast a documentary in December 2006 concluding that Danish soldiers had handed prisoners over to the United States despite knowing they would be tortured and transferred to the Guantanamo Bay detention camp. The accusation was serious; if true it would mean that not only did Danish units commit a crime but that – according to the filmmakers – the political and military leadership did nothing to prevent the transfers despite being in the know (*Information* 2010). Since there were no hard facts to corroborate the programme's conclusions, it all boiled down to a question of credibility (Mollerup 2007, Hoffmann 2010). In the government's 24-page response, all accusations presented in the programme were rejected (Forsvarsministeriet 2006). Nevertheless, the Prime Minister's main attack was on filmmakers' credibility. He announced his intention to launch a public investigation to uncover why the programme had been bought in the first place (DR.Nyheder 2007). This plan was dropped; criticism from the opposition parties, which so far had been loyal supporters of the government's Afghanistan policy, was strong. The fact that the film ended up earning a prize as best documentary of the year may also have played a part in the government's decision to shelve the planned investigation. Had it persevered, the government might have seen its own credibility suffer.

If the official frame of a war is undermined by facts, credibility suffers. This became acute for the German government, which continued for years to promote an image of the Bundeswehr's deployment as peaceful. Admitting that Bundeswehr soldiers were participating in aggressive actions would have weakened the government's portrayal of the armed forces as protecting the local population and assisting them in the reconstruction of the country. This official narrative lost credibility with each new reported death. The problem for the government was how to respond without losing still more credibility in the eyes of the public. The challenge became acute when dead soldiers were returned for burial. Were they fallen soldiers killed in a war, or victims of unfortunate circumstances that cropped up during a development mission? Robert Entman, an American social scientist and media expert who has contributed much to the development of framing as an analytical tool, has pointed to the need for a government to maintain congruence with dominant 'schemas' in order to ensure that a message is believed (2004: 14). If done successfully, the scope for criticism will be small. If the official frame is perceived to be incongruent with developments on the ground or with the accounts of high-credibility sources, like returning soldiers, criticism will likely gain traction and support for the war effort will slip. It should be underlined that congruence and incongruence are only terms measuring approximation or distance. The crucial question is to what? A specific answer depends on the country in question. Both the British and Danish governments' frames were uncontested. In Germany, the political leadership faced an uphill battle in its attempt to gain political acceptance for the policy chosen. The reason was the lack of a clear consensus, not so much pertaining to Afghanistan as to which security policy role Germany should choose.

Government or military spokesmen providing an official explanation and journalists reporting on casualties are both confronted with a similar credibility challenge. They must be perceived as trustworthy. If not, their message will be disbelieved no matter how close it is to the truth. The best-known example of a breakdown in trust occurred during the Vietnam War. Despite upbeat bulletins from the military and political leadership, graphic images from the battlefield horrified viewers at home (Braestrup 1985). During the Tet Offensive in 1968, an estimated 20 million viewers across America watched the operations on their television sets each evening (Thrall 2000: 31). What people saw was very different from the official version of events. What was later nicknamed 'the CNN effect' refers to the corrosive impact of alternative accounts on public support (Livingston 1997, Smith 2006: 284–9). The authorities respond to criticism from the media, from other politicians or from their own armed forces. Often this is done as a matter of balancing, for instance claiming that some presentation/analysis/discussion was biased and in need of official clarification. General William Westmoreland, in charge of the US Army during the Vietnam War, was aware of the impact public opinion had on the 'prosecution of the war' and claimed that 'the accuracy and balance of the news coverage has attained an importance almost equal to the actual combat operations' (Sharp and Westmoreland 1969: 274). The Vietnam War showed how wrong official information managers can go. By the end, despite official daily efforts 'to balance' critical reporting, two thirds of the American public said they did not consider government statements about the war to be credible (Thrall 2000: 19).

The lessons of Vietnam have affected the way casualties are reported. In Afghanistan, the use of language in official communiqués has remained similar from country to country. Nonetheless, the accounts vary in how they describe what the soldiers were doing when killed, the sort of attack they were exposed to and the character of their missions. It's no coincidence that the peaceful character of a soldier's mission or his involvement in civilian reconstruction was mentioned more frequently in official German communiqués than in those issued by authorities in the two other countries, where courage and combat skills loomed larger. It is by such comparison that dominant national schemas can be outlined along with the strategies adopted by politicians when they have to convey facts that break with them.

In the few examples given above, attention has been directed to official frame making – how governments would like their populations to interpret what is happening in Afghanistan and why men and women are dying there. The government and military leadership have enjoyed an almost complete monopoly in informing the public about losses; in journalistic sourcing, public officials were lopsidedly overrepresented. There have been notable exceptions, especially when friendly fire was the cause of death or miscalculations led to large-scale civilian losses (Fischer 2010). Yet, on the whole, official control over information made it possible to tailor how losses were conveyed to the public. Facts were vetted, and needless to say, all approved versions have adhered to the official framing of the war as a necessary response to threats emanating from Afghanistan. With the exit date of 2014 approaching, governments and military leaders in all three countries would also assert that preparations for a smooth transfer of security responsibility to local Afghan forces were running according to schedule.

Alternative Versions

The official version would have been challenged by any independently verifiable alternative versions of the war. In the immediate post-invasion phase, Afghanistan was 'the most reported-from country on earth' (Zelizer 2010: 269). As the war lingered on, most journalists left. As a consequence, the official disseminators of information have come to enjoy a near monopoly. This is most clearly seen in the way casualty news was conveyed. Those best able to question the official accounts were other soldiers, and they have indeed been attentive, reacting publicly when they believe events were incorrectly portrayed. In most cases, official communiqués were too brief to contain much that could be disputed, but there have been cases when the death of a soldier became front-page material in part because of information given to the press by fellow soldiers. Often their accounts were the only alternative version of what has happened. How the men and women in Afghanistan have described the experience of losing their comrades has been an important source for this book.

Ideally, newspapers and television should provide alternative versions of events; they should be the ones balancing official news releases and providing the reader/viewer with the impartiality needed to form a sound independent view. In Afghanistan, opportunities for investigative journalism have been limited. Security was the main reason; few journalists travelled on their own. The Taliban did not distinguish between Western soldiers and journalists, and since the war began 23 journalists have been killed (Committee 2012). Most journalists reporting from Afghanistan were embedded with the troops, especially when outside of the capital. This practice has been criticised as making independent reporting close to impossible (Mans et al. 2008). A reminder that this is not necessarily the case can be found in Margaret Thatcher's comments on news reports from journalists embedded with British troops during the Falklands War: 'I also became very unhappy at the attempted 'even-handedness' of some of the comment, and the chilling use of the third-person-talk of 'the British' and 'the Argentinians' on our news programmes' (Thatcher 1993: 181). Thatcher's views notwithstanding, embedding is perceived by many observers as a hindrance to objective reporting because journalists automatically develop strong emotional bonds to the soldiers protecting them. A famous example of such bonding was provided by John Simpson, the star BBC war reporter, who entered Kabul as part of a UK military unit and reported: 'It felt extraordinarily exhilarating to be liberating a city which had suffered so much under a cruel and stifling regime' (BBC 2001). Quite apart from the news value of Simpson's report, his early entry into Kabul added to the credibility of analyses he would provide later: he had been there, had seen what happened. Questioning his observations was difficult without access to an alternative account of events. At the time, there was none.

The Simpson case illustrates a fundamental aspect of embedding: without the troops, journalists would be unable to travel. By extension that means that journalists go where the military wants them to go. Gina Cavallaro, an embedded correspondent for the US *Army Times*, has concluded: 'They're relying more on the military to get them where they want to go, and as a result, the military is getting smarter about getting its own story told' (Eisele 2005). General Rupert Smith, one of the few to write about the importance of the media from a military perspective, is quite open about this in his assessment of the media as an integral part of how modern armed forces 'fight and operate amongst the people' (2006: 284).

The military's increasing shrewdness about the media can easily lead to an overemphasis on military success stories. No journalist, as far as can be ascertained, has been embedded with troops beaten in battle and forced to retreat by the Taliban. The importance of good news from the front line to the support of the war effort is no secret. The British official strategy paper on how to ensure support recommends embedding journalists with troops 'when worthwhile stories can be anticipated (the Musa Qaleh offensive in Afghanistan is a good example) …' (Davies et al. 2008: 13). Especially in Germany – which has embedded fewer journalists than the two other countries, to judge from the number of articles coming out of Afghanistan – embedding is often derided as a military propaganda tool (Sponsel 2003). However, even unembedded journalists are not immune to taking the official line. Just after Kabul was liberated, when the city was swarming with journalists, the execution of war prisoners was reported as 'vengeance for Taliban brutality' (Zelizer 2010). The fact that most had been hideously tortured before being killed was barely mentioned.

Apart from letting journalists travel to areas that would have been unreachable without military protection, the obvious benefits of embedding include allowing journalists to see the war from the soldiers' point of view and to ask soldiers for their impressions (Farrell 2010). Very often those views and experiences contradict the official frame, focusing instead on the dangers faced by the troops (Chivers 2011). Nevertheless, the policy of embedding journalists has been a clear media victory for the governments and armed forces (Lindner 2008). It can in any case result in a transfer of focus away from events to the journalist and how he experiences hardship and danger. Danish TV reporter Rasmus Tantholdt, for example, was visiting a Danish unit in the field in late 2007 when the Taliban launched an attack. The unit received air support and the Taliban was, in all likelihood, bombed to smithereens. The Danish soldiers cheered as the bombs were dropped. Critical voices at home claimed that this was a cruel gesture, but the objection found no support in public opinion or among the political parties. The terrified face of Tantholdt during the attack effectively halted all attempt to launch a debate over what critics perceived as unacceptable behaviour by the Danish soldiers (Roed 2007). The authenticity of Simpson's and Tantholdt's experiences turned the reporters themselves into the main story (Luhmann 2004: 24–31). Their emotional experiences were at least as important as what actually occurred. This was not necessarily negative from an official vantage point.[6] The government and the military leadership were assured a steady stream of war-zone reporting focused on the experiences of their soldiers in their section of Afghanistan.

Because of embedding, the official version of events in Afghanistan has rarely been challenged by the media. In only a few cases have alternative versions been published at the same time that official information has been made available. The German Kunduz affair was exceptional in that media reports about a bombardment forced the German Ministry of Defence and the Bundeswehr to retract the initial version of events and launch a full investigation (*Spiegel* 2009). It was revealed that up to 100 civilians were killed when a German plane had blown up an overturned oil tanker to prevent the Taliban from retrieving the cargo. The credibility of the Bundeswehr suffered as a result. For a long period in Britain, the question of whether troops had the necessary protective equipment to combat the Taliban was a recurrent topic in parliamentary debates and in the media whenever casualties surged. As in Germany, the official version was challenged, and in both countries the question of responsibility for casualties coloured both the questions and the official replies.

Kunduz was a singular event, for reasons that will be explained. In Britain, the equipment debate popped up at intervals. Another difference was that while the media and politicians in Germany challenged the initial account of Kunduz – for good reason, as it turned out – in Britain the official accounts of combat deaths were rarely disputed. British blog posts by present and former soldiers respond to casualties with expressions of sadness and despair, no differently from how their Danish and German colleagues express themselves. Such blogs often provide unfiltered accounts of events, supplementing the anodyne official reports. They are not uncontroversial, and the authorities have kept a sharp

6 For an analysis of how embedding biased news reporting in the Iraq War, see Kull et al. 2003–4.

eye on them. Yet, as far as can be ascertained, the objections raised have never challenged the official description of a fellow soldier's death. Official reactions restricting publication on websites centred on potential image damage rather than pay problems or inadequate equipment. A flurry of videos posted on YouTube of the 2006 deployment was certainly unwelcome to British authorities; most of the videos reflected the atmosphere immediately following combat and thus gave a more macho impression than the Ministry of Defence found to its liking. The videos were later removed from YouTube. Their release in the first place may have been one reason the UK's MoD issued a new set of rules in August 2007 governing soldiers' use of blogs, cameras and electronic media. Similarly, the Danish Defence Command has issued guidelines to limit the information shared by Danish soldiers on Facebook (Fonseca 2009). Despite the lack of any verifiable data on the effects of these measures, blogging seems not to have been radically affected. Blogs are often started and abandoned abruptly.[7] The exceptions are memorial sites run by the British MoD and memorial sites set up by some Danish contingents. None of these contain any criticism of the official communiqués announcing soldier fatalities, let alone of the war itself. They can be contrasted with German blogs and independent websites maintained for brief periods of the war. In those, the authorities were condemned for neglecting to honour dead soldiers by attending their funerals or expressing official sorrow over the loss of life.

Domestic Criticism

Criticism of the war and concern over casualty rates have persisted throughout the war in all three of the countries examined here, but credible alternative views have been limited and, as will be shown, have often come from soldiers in the field. In Germany, antagonism did lead to a change in the official line on the war, but it was never strong enough to threaten the official decision to keep troops in Afghanistan. To have achieved that, the critics would have had to connect with anti-war groupings in the rest of society, and that never happened. Even soldiers in the field with a critical perspective on the war seem to have little but scorn for war opponents at home, and the feeling is most likely reciprocated.

In short, at no time has anti-war sentiment in the population been sufficient to threaten the governments' official war views. In all three countries, parliamentary elections have been held with the incumbents re-elected: Britain in 2005, Germany in 2009, and Denmark in 2005, 2007 and 2009. Even when there was a change of political colour, the incoming government continued its predecessor's policy on Afghanistan. In Britain, the Liberal Democrats had since 2009 demanded a swift withdrawal; when they joined the Conservatives in a governing coalition in 2010 this view was quietly shelved.

7 The German Soldatenglück.de, the British defenceoftherealm.blogspot.com and the Danish krigeren.dk are among the few that have been maintained since the start of the war.

Nevertheless, neither political continuity nor the absence of strong political opposition should be equated with permanence in popular support. A notable lack of attention from the media and the public did not signal support for the governments' plan to maintain and increase the number of troops deployed until a withdrawal date had been agreed by all the members of the Western coalition. That popular patience was wearing thin was evident to all political leaders: Prime Minister Cameron's announcement of 2014 as the withdrawal date can best be characterised as a pre-emptive strike aimed at domestic opinion. No developments on the ground in Afghanistan had indicated that stability would return to the country by that date. Among those who have expressed the strongest doubts about the feasibility of an orderly transfer of security responsibility to local Afghan forces are the men and women stationed there.

Cameron's announcement came at a time when war weariness was increasing in the British public, and the same trend could be observed in Germany and Denmark. In Britain and Denmark, the prime ministers went on air claiming that defeatism at home encouraged the Taliban to increase attacks on Western forces (Channel 4 2010, *Jyllands-Posten* 2010b). The statements came as part of government efforts to drum up support. The original justification, that war was necessary to prevent terrorist attacks at home, was repeated. It was no longer the sole or even the main justification for continued deployment, however. In the official rhetoric, development and assistance came to play a stronger role. The number of wells drilled and girls' schools opened would be cited in political speeches and in the official information brochures published by the British and Danish governments and made available on the Internet and in public libraries (FCO and DFID 2005, Bundesregierung 2008, UD and FD 2008, 2010). In Germany, such factors had been the official rationale for deployment from the beginning. While this strategy has not failed outright, neither has it seemed to pay off. Development issues have rarely caught the media's attention beyond perhaps a few lines in articles and on the evening news referring to concrete progress. Not even when soldiers wrote books referring to civilian reconstruction programmes and the significant improvements they often represented to local inhabitants would such matters rise to prominence in the media. In book reviews, the authors' combat experience overshadowed all else. To be sure, combat loomed large in what the soldiers had written, but reviewers tended to overlook the attention also given to economic development and the role of Western development aid in addressing terrorism. Governments, too, overlooked this aspect; they could have used the soldiers' observations in support of their own policy, but did not.

With the withdrawal of coalition troops approaching in 2014, official attention shifted towards the transfer of security control to local Afghan forces. The emphasis given at the start of the war to achieving democracy, equal rights, stability and prosperity disappeared from official rhetoric. The men and women who left home to serve in Afghanistan have been less quick to forget the original rationale. Concerns over what may happen after they withdraw emerged in many books and accounts (Birkeland 2010: 77–84, Harnden 2011: 489–90, *Süddeutsche Zeitung* 2012). The gap between the official line (that everything was essentially on schedule) and contrary views from soldiers in Afghanistan and independent security scholars left the governments in a weak position (Hoffman 2012, Matern 2012, Tuxen 2012). The question is why criticism has not been more widespread.

To answer that, it is necessary to gauge the extent and origins of the criticism expressed in parliaments and the media. Within the large political parties, a few lone voices opposing the war have been heard in all the three countries.[8] On the parliamentary fringe, two parties opposed the war from the beginning: the Left [die Linke], a successor to the former East German communist party, and the Red-Green Alliance [Enhedslisten] in Denmark. Their best election results, respectively, were close to 7 per cent for the Alliance (2011) and 12 per cent for the Left Party (2009). Their gains were due more to the economic crisis and rising unemployment than the war. Of the two, only the Left Party has consistently exploited its role as sole opponent to the war to gain attention. As will be discussed in greater detail below, the Left Party's ability to dominate its side of the debate has in no small measure been due to the reluctance of other parties to go against the official line.

The two fringe parties' efforts were echoed in the writings and comments of prominent intellectuals like the Danish writer Carsten Jensen and the German Günther Grass; in Britain, Harold Pinter gained attention as an outspoken critic (*Spiegel* 2001, Chrisafis and Tilden 2003, Berlin-gegen-Krieg.de 2009, Jensen 2010). Their impact was at best negligible. Compared with the war in Iraq, where intellectuals and politicians together played a significant part in mobilising opposition, no extra-parliamentary protests have dented the governments' Afghanistan policies. Public protests have focused on other conflicts, such as the Israeli war against Hezbollah (2006), the bombing of Gaza (2008) and above all the Iraq War. Placards and slogans demanding withdrawal from Afghanistan could be seen and heard during demonstrations, but this was not a war that drew masses into the streets. The lack of attention was also felt among decision makers, with the British Chief of the Defence Staff Sir Jock Stirrup complaining that 'Iraq was burning all the political oxygen' (Stirrup 2009).

The prevailing consensus in all the three countries meant that several war issues went unremarked upon by either the government or members of parliament. Most notable has been the lack of debate about the war's economic costs, including the sums to be spent on rehabilitating soldiers for years to come. At a time of economic crisis, this may be a cause for surprise. Britain is an exception in that an assessment of the total costs of the deployment was published by the House of Commons Foreign Affairs Committee; even there, however, the sums required for treating veterans was left uncalculated beyond an admission that the final price

8 In Britain, the Labour MPs Jeremy Corbin and Kate Hoey played this role.

would be pushed significantly upwards (2011: 19). No debate ensued, possibly because it would be seen as a lack of commitment to the troops.[9] In the two other countries, no official estimate was made public. In Germany, a group of researchers presented a detailed calculation of the costs, including future medical treatment (Brück et al. 2010). Despite the high final sum arrived at, the government made no response. In Denmark in early 2012, journalists added the sums allocated to the war by the Ministries of Defence and Foreign Affairs to try to come up with a rough idea of the total cost (*Information* 2012b). Without reliable data, estimating future medical needs for veterans was impossible (*Information* 2012c, 2012d). Nevertheless, the journalists' conclusion was solid enough to justify asking whether Denmark had paid too high a price for the results achieved in Afghanistan (*Information* 2012a). The Minister of Defence responded by repeating the same argument made in 2001: that Denmark had no alternative but to participate and that 'the war is worth the price' (*Information* 2012b).

A far greater threat to government war policies than a debate over money spent is criticism from senior officers. As pointed out above, if the elite cannot agree on the need to fight a war, the government will lose public credibility and support for the war will suffer. In the three countries examined here there have been some notable cases of elite criticism of the Afghanistan war effort, not from government members but from military leaders. Most commonplace have been complaints over inadequate equipment. More serious were accusations that some countries had purposely assumed a distinctly low-risk approach, leaving the dangerous tasks to others. In late 2008, ISAF General Hans-Lothar Domröse openly stated that the coalition was riddled with animosity between 'those countries that mainly fought and those mainly committed to well drilling', clearly implying that Germany belonged to the latter (Klein 2008). Although the statement echoed complaints made by the Americans, the British and the Danes, it failed to have any impact on German public debate because there was no support in the Bundestag or the population for a more offensive German role. More serious were the general's reported doubts that the war could not be won (Klein 2008). Although repeated in many German newspapers, these drew no comment from the government. Nor did anybody in the Bundestag press the defence minister for a comment. The public response to Domröse's assessment was non-existent. A journalist concluded that 'the power of adaptation' to the war was sufficiently strong to put a lid on public response (Blechschmidt 2008).

Consensus has reduced pressure on the three governments to constantly justify their deployments to Afghanistan. By contrast, the Danish and British governments have faced rigorous questioning about their participation in the Iraq War. The relative harmony on Afghanistan policy has not, in any case, relieved them of a duty to issue formal responses to combat deaths. These have come in the form of tributes when coffins have been repatriated and at funerals. Neither

9 Hence the careful wording of MP Nick Harvey when Afghanistan was debated in November 2011 (Hansard 2011).

occasion lends itself to debate; few critics would exploit the death of a soldier to attack the government. The exceptions, to be shown in greater detail below, are when casualty numbers and individual deaths are cited in parliamentary debates, although in those cases the government is never pressured to withdraw. Of similar nature was a statement by the German Federal Armed Forces Association [Deutscher Bundeswehrverband] in October 2008, claiming that the government bore responsibility for the unnecessary deaths of two men and the maiming of a third in a suicide attack (*Süddeutsche Zeitung* 2008). Particularly offensive from the government's viewpoint was the Association's assertion that the attack could have been prevented if civilian reconstruction had been prioritised more highly. The government did not respond in words, but a shift of economic priorities became evident in subsequent years, with civilian projects receiving a greater funding share.

Accusations of government responsibility for combat deaths in an unnecessary war and demands for immediate troop withdrawal have been rare in the case of Afghanistan. By comparison, the only time such an accusation/demand was made in the Danish Parliament was after a Danish soldier had been killed in Iraq in May 2007. The head of the Socialist People's Party said the death was the government's direct responsibility (Klarskov and Svane 2007). A handful of politicians followed up this accusation with calls for an immediate Danish withdrawal – calls that went largely unanswered. During the Afghanistan war, debate unfolded differently. Not only the government, but also soldiers would often respond to criticism. For instance, when the leader of Germany's Left party, Oskar Lafontaine, claimed that the Bundeswehr in Afghanistan was promoting terrorism, the men on the ground spoke out (*Welt* 2007). A soldier who had survived a Taliban attack that killed three of his comrades pointed to the statement as a particularly wicked form of backstabbing of the men on duty (Schmuda 2009).

Lafontaine and his party were a lone source of continuous criticism, posing no threat to the German authorities. Yet Germany's government, like those in Denmark and Britain, had somehow to justify the lives lost in Afghanistan. In all three countries, ministers argued that the importance of casualties should not be exaggerated. This tactic was clearest in Britain, where Bob Ainsworth, Secretary of State for Defence, warned against letting the number of dead distract attention from the overall Afghan strategy of transferring power to local security forces (Ainsworth 2009). In Germany, Minister of Defence de Maizière stated that doubts about the war were permitted, but must be overcome (Meyer 2011). His message was the stronger because he delivered it in church, before the coffins of three dead soldiers, and because it was one of the few statements by a German government member covered by the press during a funeral ceremony. Danish Prime Minister Fogh Rasmussen has used a funeral ceremony in similar manner to whip up support for the soldiers in Afghanistan, strongly implying that a failure to do so would render meaningless the sacrifices already made (Steenbuch 2008).

Such statements indicate if not an outright shift in the main official framing of the war, then certainly a broadening. An official UK information brochure published in 2005 emphasised assistance, aid and reconstruction (FCO and DFID 2005). Danish and German informational material was much the same (Bundesregierung 2008, UD and FD 2008). The photos in all such publications resemble each other in that they show numerous happy, clean and well-fed children flocking around the men on patrol. The German brochure was almost cleansed of military imagery (Bundesregierung 2008). In the only photo of a Bundeswehr man carrying a rifle, the muzzle is pointed downwards. One journalist wrote that the brochure portrayed the Bundeswehr as a lightly armed security force guarding German-financed development projects (Erhart and Kaestner 2008). Combat and death were missing not only from the photos but from the text as well. The earliest official publications contained a few references to the necessity of keeping soldiers on the ground until the security situation improved. With the increase in casualties and the receding prospect of victory, information strategies have changed. In a 2010 brochure published by the Danish ministries of defence and foreign affairs – entitled *Denmark in Afghanistan – Why, Where, How Long?* – the security situation is addressed in greater detail and in more direct terms than in previous information leaflets (Dueholm 2010). Such brochures have been only part of the overall information picture. For most readers, the emphasis on progress would automatically be countered by media images of coffins and mourning relatives. And these images increased in frequency throughout the war.

Chapter Outline

The framing of fallen soldiers by political leaders has not been a static phenomenon; likewise, the media debates and the receptions and commemorations when coffins return have changed as time has passed and lessons have been learned. The politicians have adapted their behaviour visibly in the course of the war in response to public pressure. Why and how will be addressed in Chapter 2, where the political backdrop for the deployments is outlined. In all three countries examined, the Afghanistan mission was presented as peacekeeping. Whereas the Danish and the British governments soon shifted rhetoric and underlined the need for combat troops, the German government resisted attempts to frame its deployment as anything but a development and assistance mission. When dead soldiers started to be sent back home, the official German line was one of no-response. In Denmark the opposite approach was chosen, with the government actively using Danish participation in the war to further its transformation of security policy. The dead were lauded as admirable Danes. As such, they were beyond political attack yet at the same time potent political instruments in the government's strategy. In Britain, too, the dead soldiers came to play a highly political role. They were referred to constantly in debates about the conduct of the war and the government's funding of troops in the field.

In Chapter 3, the repatriation and burial of dead soldiers is analysed. The ceremonies lend insight to the political and public responses in the three countries.

The men and women themselves, the dead soldiers, often had strong views on how such ceremonies should be conducted. A few of those killed have achieved fame, pushed into the public consciousness largely by their fellow soldiers. What this status means will be outlined.

In the course of the war, demands for memorials started to emerge from the soldiers themselves and from the families of those killed. The three countries have responded very differently. In Britain, the war dead of Afghanistan have been included as part of the country's long record of men and women lost in previous wars. In Denmark, a memorial has been erected which clearly expresses the government's political line on the war. In Germany, by contrast, debate over whether the country should have a monument to the war dead was more vehement and bitter. The resulting monument is in many ways an unfortunate compromise. This is the topic of Chapter 4.

Finally, in the conclusion, the significance of war dead in contemporary society will be discussed. In all three countries, casualties in Afghanistan have had an impact on national identity. Both Denmark and Germany have traditionally been regarded as conflict averse, not only by international observers but by their own populations. Afghanistan has affected this perception.

What binds these chapters together, apart from the fallen soldiers, is the focus on how the war has been framed by different parties: the government and military leaderships, the media, and the soldiers in Afghanistan. The emphasis on framing does not lend itself to the sort of clear-cut answers often found in quantitative casualty-sensitivity studies. On the other hand, it does become possible to say something about the role that combat deaths play in strategic culture in the countries analysed. Many may balk at this: 'Most students of national security accord pride of place to material forces that define the balance of power between states. They have no patience for intangibles like culture', the American political scientist Peter Katzenstein concluded (Katzenstein 1998). In this book, attention will be paid to such intangibles. It may sound contradictory but the dead men and women sent back in coffins have been, in certain ways and to varying degrees, intangible. That's especially the case in Denmark, where the government has assigned them heroic status, and in Germany, where leaders have feared that too much attention would undermine public and political support. This book is an attempt to explain why this has been so, and how things changed in the course of the war.

Chapter 2
Sending Off the Living

Soldiers returning in coffins to Denmark, Germany and the UK have been received in a style consistent with their send-off to Afghanistan. Each government's attitude to the return of coffins has reflected its framing of the war and its original rationale for deployment. National differences were visible from the start as each country struggled in its own way to balance civilian assistance with security concerns. Indeed, Danish, German and British soldiers quickly found themselves on different missions. Not only was each national force assigned to a separate region of Afghanistan. It also carried a distinct mandate from home, reflecting how its particular government had framed the war. When casualties began arriving home, the three governments were forced to justify the deployments in new terms: they had to convince their populations that the war was worth the cost. That is the subject of the present chapter.

After the terror attacks in the United States on 11 September 2011, the countries analysed here all perceived the same type of threat emanating from Afghanistan. 'Further attacks were expected any day. ... widespread anxiety could anytime give vent to panic', as Frank-Walter Steinmeier, head of the Chancellery in Berlin, has recalled (2011). New York and Washington had shown with all clarity that nobody was safe. In the preceding years many Western European countries had suffered attacks by Islamist terrorists, but never of such magnitude and in front of running cameras. Politicians and security experts immediately discussed possible links between Al Qaeda and jihadist cells in Western Europe, and it was soon discovered that a great deal of planning before the 9/11 attacks had been done in Hamburg (National Commission 2004: ch. 5).

If people in the West were to be protected, the Taliban regime that had sheltered Al Qaeda in Afghanistan would have to be rooted out. Each intervening country would therefore frame its deployment as a self-defence mission – as an act of necessity, not of choice. Referring to British citizens killed in the Twin Towers attacks, Tony Blair declared that not only the US but also Britain had been attacked and would retaliate (Blair 2001d). In a speech to Parliament he expressed it thus: 'So this military action we are undertaking is not just for a just cause alone, though this cause is just. It is to protect our country, our people, our economy, our way of life ... if attacked, we will respond. We will defend ourselves' (Blair 2001a). Blair may have been the first to link the war against the Taliban to the possibility of Islamist attacks at home. If the Al Qaeda command centre in Kabul were annihilated, he suggested, terrorist activities would subside in Britain as well (BBC 2002). Political leaders in Denmark and Germany could soon be heard making similar assumptions. Through frequent repetition, the link to security at home would emerge as a key reason why Western soldiers were fighting in Afghanistan. When Gordon Brown unveiled his new strategy for the British deployment in 2009, he stated that it would 'mean more security on the streets of Britain' (Stratton 2009). The following year, after ousting Brown at the polls, the Conservative-led coalition repeated his rationale almost verbatim (MoD 2011).

Soon after the 9/11 attacks, German Chancellor Gerhard Schröder declared in the Bundestag that 'a war against the civilised world' had been declared, and he could not exclude the possibility that Germany's contribution to the response 'might also involve military assistance' (Schröder 2001: 18302C). In his view, however, military means should play at most a secondary role to development aid and assistance. Even this rather timid admittance caused protests; the potential deployment of German troops outside was perceived by peace activists and politicians on the left as toadying to the United States. One slogan that appeared was: 'For the Fatherland and Bush, at the feet of the Hindu Kush' (Pau 2003). Starting with the first German troop send-off, the basic thrust of criticism was double edged: that the deployment was done to please the US, and that it was an expression of a re-emerging German militarism (Henken 2006, Lüders 2006). Whereas right-of-centre parties supported the deployment as a sign that German security policies were becoming more like those of other Western countries or normalising to use the German expression, the parties of the left wanted Germany to continue its traditional emphasis on peaceful negotiations and refrain from military action. This more dovish attitude, widespread even in Schröder's Social Democratic Party, was often tinged with a degree of anti-Americanism not seen in the other countries examined here (Diner 2002, Freund 2008). Divided internally and opposed from the left by the Party of Democratic Socialism, the governing coalition of Social Democrats and Greens could not be sure it would receive the backing in the Bundestag that it needed for deployment. To force through a Yes vote when the issue was debated in November 2001, the Chancellor said he would resign should he fail to get enough votes. A No vote in the parliament, he argued, would irrevocably damage Germany's credibility as a reliable ally (Bundestag 2010a). In the end, 336 members supported him, only two more than the minimum required (Bundestag 2010a). A member of Alliance '90/The Greens, traditionally a strongly pacifist party, defended her own Yes vote by underlining that the Bundeswehr's mission would be of a humanitarian nature (Bundestag 2001: 1968–69). Her assumption seemed fairly well grounded: in the debate leading up to the vote, the government had said nothing to suggest otherwise.

In retrospect, it would be easy to accuse the German government of misleading MPs over the true nature of the deployment. Yet at the time, it was far from clear what the United States would require from its allies. The Danish parliament, too, debated its military contribution with few facts on the table. As a result, the Danish mandate for the mission was put in almost rudimentary terms: apprehend those responsible for the terror attacks; close down the terrorist training camps; and remove the Taliban regime (Dalgaard-Nielsen 2008: 33). Although no time limit was given at the time, there was nothing to suggest a long-term commitment would be required. In fact, developments were so quick that some Danish MPs questioned whether there would be any job waiting for the Danish soldiers who were then assembling for departure (*Jyllands-Posten* 2001c). Nobody expected the war to last much longer than it would take to oust the Taliban from Kabul. In Denmark, some MPs wanted to leave combat to other countries (read: the US and Britain) and take responsibility instead for mine clearance and peacekeeping (Gram et al. 2002). With an end to the fighting apparently already in sight, they thought it would be futile to send Danish troops.

Still, reconstruction and civilian assistance played an insignificant role in Denmark's parliamentary debate at the time – not because of any preference for a military solution, but because so little was known about the needs on the ground in Afghanistan.

Peacekeeping or War?

The first troops sent to Afghanistan were expected to fill the power vacuum left by the fleeing Taliban. Peace was regarded as achievable within the near future. In Germany, the newspapers referred to continuing fighting as the final skirmishes against fundamentalist die-hards (Berliner Zeitung 2001). Similar views were expressed in the Danish and British media as well (Dutter 2001, *Jyllands-Posten* 2001a).

The lack of in-depth knowledge about Afghanistan and the cheers of Kabul residents that greeted coalition forces certainly helped account for the public optimism; such factors even influenced official descriptions of what the soldiers were being sent to achieve. Two disparate approaches soon emerged, one in Germany emphasising humanitarian remedies and the other in the United States and Britain advocating a strong military reply. The disparity can be traced back to the United Nations and the two resolutions passed after the September 11 attacks on New York and Washington, DC. One provided a mandate for the actual war against the Taliban labelled Operation Enduring Freedom starting 7 October 2001.[1] The other, passed by the UN on 20 December the same year, led to the establishment of ISAF (the International Security Assistance Force). Initially, ISAF was given responsibility for security in Kabul. From 2003 the mandate was changed to include responsibility for security throughout the country, and command over ISAF was transferred to NATO. The change was a belated move reflecting realities on the ground. It was a sign that the war had entered a new, more serious phase and that a swift end to combat activities was not in sight. Taliban activities spread to regions that had been relatively calm, including the German-controlled north, which until then had been far more peaceful than the southern areas where Danish

1 Originally it was named Infinite Justice, but the religious connotations caused an outcry in many Muslim countries, forcing the Bush Administration to change the name to Enduring Freedom.

and British forces operated. The approach of hostile elements had little if any impact on the official German line as expressed by the government in its various brochures on Afghanistan and presentations to the Bundestag. Germany's role was framed as distinct from what the other countries were doing; the government in Berlin emphasised its role in co-ordinating international assistance by hosting four key international conferences on stabilisation and rebuilding (Schlagintweit 2005, Overhaus 2009: 269). The first of these took place in late 2001. It ended with a plan for a stepwise transition towards democracy and stability in Afghanistan (Auswärtiges Amt 2001). However, before it had been agreed, cracks emerged between the participating countries over what role Western forces should play. Those cracks received little attention in the German media. When conflicts were noted, they were often framed as disagreements within the alliance over a complex understanding of security with some countries putting too much confidence in military might (Chauvistré 2001a, 2001b). Germany's perception that it had a more sophisticated approach to security stemmed from its post-1945 tradition of peaceful conflict resolution (Maull 2007). A key flaw in the desire to rank Germany's approach as superior was the simple fact that the British and Danish governments had by no means forgotten civilian assistance.

There were, nevertheless, real differences in approach that would have a lasting impact on what the three governments allowed their soldiers to do on arrival. Most countries sending troops wanted the United Nations to agree on a mandate consistent with Chapter VII of the UN charter. That would mean troops could resort to arms for purposes other than self-defence alone. Germany opposed this, while Britain and Denmark were strongly in support. When the Security Council finally passed the resolution leading to the establishment of ISAF, the mandate was based on Chapter VII and allowed soldiers to 'take all measures necessary to fulfil its mandate' (Security Council 2001: 2). In practice it would be up to each country to decide exactly what measures to authorise. From the start, in consequence, the three countries gave their soldiers widely different instruments.

The conflicting national views were also reflected in the number of soldiers that leaders of the three nations believed would be required to uphold the mandate. Britain was alone in favouring a substantial deployment from the outset. It insisted that a token force would be unable to maintain security and search for Taliban remnants at the same time. The British view met opposition from the newly formed Afghan government, and in particular from representatives of the Northern Alliance, which had been crucial in ousting the Taliban; they wanted the Western presence restricted to a few, clearly delineated areas such as Kabul. Germany came out in full support of the Afghan position. During a Security Council session in December 2001, Germany clashed with Britain and the United States. One difference of opinion centred on how the UN's ISAF force should be led. The United States and Britain wanted ISAF subordinate to the US-led Operation Enduring Freedom, with its emphasis on military forces geared for combat. Germany and France both objected. Germany went further than had been expected and demanded that any military participation be dependent on a strict distinction between ISAF and Enduring Freedom. The crux was Germany's political need to maintain a clear divide between US–UK military operations aimed at routing the Taliban and ISAF's primary objective of civilian reconstruction. When the UN Security Council passed a resolution towards the end of December 2001, it left open the relationship between ISAF and Enduring Freedom (Security Council 2001).

In Germany, humanitarian assistance, reconstruction and development quickly became buzzwords to justify deployment. Chancellors and Foreign Ministers have uttered them throughout the war whenever Afghanistan has come up for debate. Not everybody believed this was the right strategy; isolated voices among the conservative Christian Democrats could be heard, including that of Friedrich Merz. He claimed that the humanitarian emphasis was insufficient. As Merz put it, Germany would have to do more than send military transport planes to 'deliver woollen blankets to the Afghans' (Bundestag 2001: 20826). Underlying his criticism was a basic question: why should the Bundeswehr undertake an operation that could be done just as well by a non-governmental organisation? But the government declined to answer, and nobody at either end of the political spectrum asked whether the Bundeswehr would be involved in actual combat. Schröder's statement in the initial parliamentary debate that Germany's contribution 'might also involve military assistance' lacked concrete detail, and nobody in the Bundestag probed the issue further. The members who debated seemed to take for granted that German soldiers would be sent to areas already at peace, where the local inhabitants eagerly awaited Western assistance in their transition to stability and democracy. Televised images of people in Kabul cheering Western forces as they liberated the city may well have contributed to the assumption that German troops would face light duty. It was in any case a comforting thought, because it accorded with the basic tenets of German strategic culture.

Britain's approach represented a significant contrast, if not fully to the extent that Germany's media suggested at the time of the 2001 conference. The House of Commons did indeed debate the proper balance between military deployment and civilian assistance. Its emphasis shifted quickly away from that of the Bundestag, however. Both in and out of Parliament, Tony Blair's early statements and speeches focused on the military aspects of the job ahead (Blair 2001b, 2001c). The need to oust the Taliban regime in Kabul and capture its leadership was given top priority because otherwise civilian reconstruction could not be pursued. At the time, no one challenged Blair's military bias or demanded to redress the balance in favour of civilian assistance. In fact, to judge from parliamentary debates in the aftermath of the decision to deploy troops, opinion seemed clearly against any shift in the balance. One MP stated, without encountering opposition, that Britain should stay out of civilian assistance and restrict itself to military input: 'In the past, we have expressed reservations about being involved in what has been described as "nation building", especially as we have been protagonists in the campaign so far' (Ancram 2001: 853). That squared with a speech given days earlier by the Chief of the Defence Staff, General Michael Boyce, who had said such activity was better left to militarily 'less capable nations' (Boyce 2002: 3).

Boyce's views found little support from British soldiers on the ground. None of them have written condescendingly about reconstruction efforts. While some have bemoaned a lack of co-ordination between military and development organisations, they have generally endorsed aid as a key component in solving security problems for both the inhabitants of Afghanistan and the Western forces operating there (Docherty 2007: 55–76).

More importantly, Boyce's desire to concentrate on military affairs and do without civilian assistance ran counter to the Prime Minister's comprehensive strategic outlook. Like the leaders of Germany and Denmark, Blair saw in Afghanistan a political opportunity to enhance his country's security profile while emphasising good intentions. In an April 1999 speech, two years after his election, he had outlined a foreign policy doctrine that opened the way for armed interventions to spread stability and democracy (Blair 1999). He employed that rationale for British military action in the Balkans in 1999 and Sierra Leone in 2000, and would revive it again for Iraq in 2003 (Dorman 2009). In early October 2001, when Blair announced to Parliament that a military response in Afghanistan was being planned, he said that armed might would be required for the transition to stability and democracy but that the 'humanitarian coalition to help the people of Afghanistan' was 'as vital as any military action itself' (Blair 2001b). As Blair underlined in the same speech, of course, the plans for military action were 'robust' while those for humanitarian assistance were just 'falling into place'.

When the House of Commons debated Blair's decision to deploy soldiers to Afghanistan, hardly a dissenting voice was heard. Afghanistan was clearly a source of international terrorism providing shelter to Al Qaeda. Moreover, the Taliban's human rights record was a strong argument for regime change, and the United Nations had provided the necessary legal backing for intervention. The arguments were convincingly in favour of deploying troops. Parliament expressed its support for the government's decision without voting (White 2009: 184–206).

Afghanistan as Political Opportunity

The governments in all three European countries saw Afghanistan as a political opportunity to enhance their international standing and improve relations with the United States, two goals they often conflated.

For Tony Blair, whose foreign policy doctrine complemented George W. Bush's views on how to combat security threats, the way ahead was clear. Shared perceptions had resulted in close co-operation between the UK and the United States after the First Gulf War in 1991, when the two countries had kept up military pressure on Iraq. For European countries, fighting alongside the United States was burden-sharing in practice. Throughout the history of NATO, American leaders had repeatedly accused European allies of the opposite – of burden-shedding. Such accusations had been prompted by reductions in defence spending by many NATO countries after the end of the Cold War (Dunne and Nikolaidou 2011). From a US perspective, Afghanistan became a test not only of whether the European allies had the necessary capacities to fight, but of whether they were willing to expose their soldiers to real risk. Britain and Denmark shared this perception of Afghanistan; Blair and his Danish counterpart both stated that keeping their troops at home was not an option (Blair 2001d).

US–British relations had remained close after the end of the Cold War, whereas Berlin and Washington had grown increasingly distant. According to Walter Russell Mead, a seasoned observer of US–German relations, Germany was no longer regarded as 'the best and most serious of the continental states when it came to facing hard realities and making sound strategic choices under pressure' (Mead 2004: 13). Germany's refusal to participate in the 1991 Gulf War had been a turning point and was widely derided as cowardly in Washington. The fact that the US-led coalition confronted an Iraqi army equipped with weaponry manufactured mainly in Germany had been noted in Washington (Schmiese 2000: 259–66).

The distance between the two allies had expanded with Germany's loss of strategic relevance. The country was no longer a front-line state, and the US military presence there had been scaled back since the 1980s. What had surprised the US side was the lack of any attempt by the German leadership to make up for its own decline in importance. An invitation by the first President Bush for Germany to join the United States as 'partners in leadership' failed to elicit any response from the German side (Bush 1992). What occurred instead was the growth of German self-assuredness, directed first and foremost against US security policy tutelage (Bahr 2000, Schöllgen 2004). Chancellor Schröder stated famously that the era of viewing the United States as a model had passed, and that from now on the government in Berlin would carve out 'a German way' in foreign and security policy (Sturm 2002). The German insistence on ISAF as a contrast to Operation Enduring Freedom should be seen in this light. ISAF offered the chance to contribute without being closely associated with US strategy. Furthermore, Afghanistan threatened another German priority: increased military might for the European Union. This was a pronounced priority for Schröder's Social Democrats that neither the conservative parties nor the Free Democrats questioned. EU strength was considered a means of distancing Germany from the United States (Schmidt 2011). The European approach to security threats focused on risk management, with peaceful conflict resolution and humanitarian and political assistance as key elements (Union Européenne 2003, Laïdi 2008).

The growing distance between Washington and Berlin meant reduced German influence not only bilaterally, but also in NATO (Haftendorn 2002). Afghanistan was an opportunity to shore up the German position; the Chancellor's rapid declaration of unconditional solidarity with the United States seemed to indicate that this was how he perceived it. Many members of the Bundestag had a different view, forcing the Chancellor to call for a vote of confidence. Opposition stemmed from misgivings about the nature of the deployment (Freuding 2009: 344–5).

Only after Schröder repeatedly underlined that the German mission would be geared to helping the Afghans rebuild their society did the Bundestag support deployment. The initial opposition left the government with clear limits to what was permissible. The Bundeswehr was not sent to fight; a sharp distinction was drawn between Operation Enduring Freedom, which was to become synonymous with war and civilian losses, and ISAF, which due to its peace-building profile was credited with nobler values (Mogg and Bartels 2007). The German attitude was that the counter-insurgency strategy applied by the United States, Britain and eventually Denmark put a dangerous emphasis on military strength and combat. In contrast, the so-called 'comprehensive approach' preferred by German leaders focused on engaging Afghan civilians, effectively relegating the military to the role of standing guard (Schreer 2010). Later, when the Bundeswehr discovered this to be a barrier against co-operation with troops from other countries, politicians in the conservative-led bloc tried to bridge the division of roles but made little headway (Hoyer 2007). The initial distinction became entrenched. Even when subsequent evaluations concluded in no uncertain terms that the civilian tasks entrusted to the Germans had failed (in particular the training of the Afghan police force), the Bundestag found little to debate and declined to reconsider the official emphasis on civilian projects (Dempsey 2006, Graw 2010, n-tv 2010). Only by insisting on the peaceful nature of the mission had it been possible for the government to gather support across the political spectrum and carry the vote on Germany's initial deployment. Whenever the government sought to prolong the mission, it plied the Bundestag with arguments centred on aid and development.

On the insistence of the government, the German contingent was restricted to a northern region of Afghanistan where Taliban presence was minimal (Blechschmidt 2006a, Fried 2006). The hotspots were in the south, where the Americans, the British and the Danes had charge of regional security. Any extension of the German area would have threatened the relative calm of the Bundeswehr's mission. For the same reason, in the early years of the war the government in Berlin would refuse to let German officers assume command positions that would have meant greater involvement in combat operations. In 2001, the government in Berlin turned down a US proposal for a German officer to command troops tasked with ensuring peace in Kabul (FAZ 2001). The German government claimed that the Bundeswehr had neither the manpower nor the resources required (Bruns 2002). Few within ISAF believed that to be the case; with one of the largest troop contingents in Afghanistan it seemed doubtful that Germany lacked the officers needed. A more likely explanation was government fear of how the Bundestag and the population in general would react to German military prominence in a mission whose military aspects had been consistently downplayed.

In the course of the war, this line mellowed and German officers received command posts. One reason was that the calm initially prevailing in the north did not last. Conditions deteriorated as the Taliban spread its activities. The number of Western military operations increased all over the country, and the north was no exception. The number of German casualties went from zero in 2006 to seven the following year. This increase notwithstanding, casualties were far higher in the more combat-intense areas of the country, where US, UK and Danish forces were deployed (Bird and Fairweather 2009: 36). Upholding the distinction between Germany's approach and the American–British form of counter-insurgency operations no longer made sense. Yet despite change on the ground, the official framing of the German role went unrevised. The underlying premise of parliamentary debates and speeches by government members remained that the German way of tackling security issues was separate and superior. German political leaders were keen to point out that the southern province of Helmand, where Danish and British soldiers were fighting and dying, was part of Operation Enduring Freedom. They skipped over the fact that the higher Danish and British losses were due to Helmand's proximity to Pakistan and the much higher density of Taliban members there than in any other part of the country. German soldiers in the north benefited from the willingness of their allies to take the brunt of the fighting.

Only after 2006 did the official German line gradually change. Less often was the assertion heard that Germany's approach was better at preventing violence and thus preferable (Hoyer 2007, Mogg and Bartels 2007). This did not mean the government would accept Bundeswehr redeployment to the more volatile south. As late as in 2011, the government again rejected the idea (*Die Welt* 2011). Staying out of the war zone was imperative to the German government and members of the parliament, irrespective of party colour. Despite repeated pressure from coalition members that were suffering heavy losses, Berlin refused any proposal that could expose the Bundeswehr to the fighting in the south. This prompted the allies, and then German journalists and security experts, to ask whether the Bundeswehr could ever be called upon for emergency assistance to allies fighting close to the German area of responsibility (Freuding 2009). The Parliamentary Commissioner for the Armed forces answered that, in such cases, the Bundeswehr would assist (Leersch 2007).

Denmark's departure point for participating in Afghanistan differed radically from Germany's. The Danish government openly pursued a policy of using the armed forces to punch above the country's minuscule weight in international affairs (Rasmussen 2006b). To the centre-right coalition elected in November 2001, the insistence on peaceful conflict resolution so prominent in the German debate had an odious ring. It appeared closely related to Denmark's 'footnote policy' of the 1980s, when the parliamentary majority would often insist that the government add Danish reservations and opt-outs to NATO decisions. As it's credibility as an ally sank, Denmark was increasingly sidelined internationally, according to leading Danish politicians (Ellemann-Jensen 2004). Only after 1989, with the collapse of the Soviet threat and the end of Denmark's status as a frontline NATO member, did the government make a timid attempt to reach beyond the established confines. In 1991 Denmark dispatched a naval vessel to the Persian Gulf to participate in the liberation of Kuwait. Nobody in parliament called for the vessel's return (Møller 2000). Denmark's participation in the war in the former Yugoslavia turned into a more important test case. While the Danish vessel in the Gulf did not fire a shot, Danish soldiers in ex-Yugoslavia annihilated a Serb artillery position after having been attacked. It was the first time Danish troops had been engaged in an exchange of fire since World War II. In Copenhagen, political and military leaders were far from certain how ordinary Danes would react. Would they demand an immediate withdrawal? As it happened, no. On the contrary, the media were full of expressions of pride over what the Danish unit had done.

When the Danish parliament voted in favour of sending troops to Afghanistan, opposition was limited. It may be that the pacifism of the 1980s had finally been vanquished, but another factor was that potential dissenters had little idea what Danish men and women would be asked to do. The initial contingent was small, around 40 soldiers; Denmark's Prime Minister stated that he expected requests for more soldiers in the near future and would therefore not put a limit on any future Danish contributions. Anders Fogh Rasmussen of the Liberals then won the Prime Minister's post at the head of a centre-right coalition, and one journalist concluded that he had given the US president a 'blank check' (*Jyllands-Posten* 2001b). In fact, Fogh Rasmussen largely repeated what his Social Democratic predecessor had stated in parliament a few months before (Rasmussen 2001). On Afghanistan, broad cross-partisan support was visible from the start and would remain so throughout the war. There was no profound disagreement over the balance between reconstruction and combat; the mission's dual nature never emerged as a divisive issue, as it had in Germany. It was all the better if reconstruction could be accomplished without firing a shot, the Foreign Minister said during a parliamentary debate in 2006; but the implication was that when military force was required it would be provided (Møller 2006).

What has set Danish participation in Afghanistan apart from Germany's has not been the total number of men and women deployed – Germany's 4,900 dwarfs Denmark's 700 – but rather the Danish willingness to serve in Helmand province, where the Taliban's presence has been particularly strong. Despite a deteriorating security situation, the Danish line has been to stay put. Danish perseverance despite high casualty figures has earned the country a great deal of goodwill in London and Washington. Danish Prime Minister Anders Fogh Rasmussen and President George W. Bush held eight bilateral meetings in the years 2002–2008, ending with a private visit of Fogh Rasmussen to the Bush ranch in Texas. No other Danish politician has been given a comparable red-carpet treatment (DR 2008). Finally, the choice of Danish Prime Minister Anders Fogh Rasmussen as Secretary General of NATO in 2009 was a clear sign of approval of Denmark's role in Afghanistan. British and US military attachés have often highlighted their gratitude by attending the funerals of Danish officers killed in Afghanistan (Hæren 2007). As far as can be ascertained, this has not been the case when Germany's dead have been buried. On the few occasions when the news media have been granted access to German ceremonies, there's been no mention of foreign attendance.

'At the top end of soldiering'

Many of the men and women fighting in Afghanistan have seen their deployment as an opportunity to test their skills: 'After 9/11, this war has real resonance and it is at the top end of soldiering, a real test', one British officer declared (Harnden 2011: 292). Or, as a captain in the British Corps of Royal Engineers put it: 'Everyone in the Engineers wants to blow up a bridge because we do a lot of bridge demolition in our counter-mobility training ...' (Danziger 2010: 76). Judging from the Danish and German soldiers' accounts, their attitude has been similar. The main difference between the three countries has been the extent to which their soldiers could put training into practice. This was decided by politicians at home and expressed in Rules of Engagement (RoE). These differed from country to country.

German soldiers operated under the strictest set of rules (Kornelius 2009: 42). This was due to the sharp German distinction between ISAF and Operation Enduring Freedom. Although German restrictions were not made public, they soon became evident in the number of operations from which the Bundeswehr was barred. Only rarely did German rigidity catch the public's attention or lead to debate in parliament about its wisdom. In one instance, debate arose when the Bundestag was asked to approve the transfer of Tornado fighter planes from Germany to Afghanistan for reconnaissance missions. Critics worried that the Tornados could be used for offensive operations against Taliban on the ground (Rühle 2007). The government argued vehemently that the planes were not to be used for combat purposes. Yet, when asked whether this meant the planes would stay grounded if allies came under attack, the Parliamentary Commissioner for the Armed Forces, Reinhold Robbe, said that in such cases the planes would be used as a matter of defence (Leersch 2007).

Soon thereafter, the German news media took up the possibility that American forces might benefit from the Tornados' intelligence-gathering and use it for offensive purposes. The planes took surveillance photos to reveal Taliban activity on the ground, including IED placement on routes used by Western forces. The photos were processed and transmitted to ISAF headquarters. However, since the Bundestag had barred the Bundeswehr from lending support to the US-led Operation Enduring Freedom, a cumbersome vetting procedure was instituted to decide which photos the Americans could see. It hardly mattered that the United States had its own real-time, drone-supplied information, which reduced the practical value any German contribution to zero (Kornelius 2009: 68–71).

The Tornado farce, as it became known, revealed a gap not only between Germany and most other countries in ISAF but also between the German armed forces in Afghanistan and the politicians back home. Those on the ground resented the restrictions imposed on them. Not only were they marginalised within ISAF, but they were limited in how they could respond to threats. The government insisted that the RoE should be strictly tailored to peacekeeping duties, effectively preventing the use of force in any activity but self-defence (Böhmer 2009). Co-operation between German troops and their allies involved in combat operations, such as the British and the Danes, was thus limited. The Bundeswehr wanted a mandate giving it much greater leeway to decide on the ground how to respond to a crisis. This was denied (Feldmeyer 2001). Basically, weapons could only be fired in self-defence. The absurdity of such restraint became obvious when German Special Forces attempted to arrest a local Taliban commander in 2005, and failed. The man fled. Since the forces were expressly forbidden to shoot a man running away, they had to let him go (Koelbl and Szandar 2008). Only in 2009 did the German Ministry of Defence reduce the number of rules and permit soldiers to fire to prevent an attack on themselves or allied troops (König 2010).

The Britain and Danish RoE were less strict from the outset, and were revised as conditions on the ground hardened (*Berlingske* 2008, Vardrup 2008). The official line in both countries was that such rules could not be determined once and for all, but had to be adapted to meet operational requirements (Lassen 2009: 169).[2] The restrictive policy of not opening fire unless directly attacked was perceived as a potential danger to the soldiers and was never interpreted as strictly as in the German case. German RoE specified that soldiers could resort to arms only if the possibility of collateral damage, like civilian injury or death, could be excluded. As the Taliban wore no uniforms, it was not always easy to distinguish between civilians and opposing combatants. Weapons were to be used only if an opponent showed the clear intention of attacking. The final decision rested with the soldier. In principle, British and Danish troops bore the same responsibility. But Germany's stricter RoE and the government's disinclination to accept the soldiers' assessment of the threats they faced led to a growing sense among Germans in Afghanistan that their mission was impossible.

From Accidental Deaths to Killed in Combat

Repatriating dead soldiers has put to the test the official framing. During the early years, the main causes of death were accidents, not combat. This seemed to vindicate the German view that what Afghanistan needed was primarily reconstruction and development. When the first German casualty was sent home from Afghanistan in March 2002, one observer noted that political leaders seemed relieved that the soldier had lost his life in an accident and not in action against the Taliban (Blome 2002). Had the opposite been the case, German officials would have been at a loss to reconcile his death with the general frame of the deployment as one of assistance. Perhaps because most of the deaths were unsensational, the German news media paid little attention to casualties anyway. The few newspaper reports from Afghanistan focused on camp-life routines and how the Bundeswehr was busy improving contacts with the local population. Little was done to challenge what one journalist called 'humanitarian

2 In doctoral dissertation on this topic, Robert Petersen claims that Danish troops found the RoE too restrictive (2012: 437–91). The empirical data provided are meagre.

hegemony' in the media coverage (Ross 2004). Readers and viewers were left with the impression that German troops faced no danger; indeed, the soldiers killed in combat were almost all American or British. As late as 2005, the Minister of Defence felt it necessary to remind the public that German casualties could well occur (Leersch 2005). His warning was timely. That year marked a turning point, with total coalition fatalities reaching 131, compared with 60 the preceding year (iCasualties 2013). Most were US soldiers. In 2006 Britain lost 39, compared with only five killed in the previous four years of deployment. While neither Germany nor Denmark lost any members of their forces in 2006, in 2007 Germany lost seven and Denmark six.

Table 2.1 Annual casualty rates by nationality, 2001–2013

Year	Britain	Denmark	Germany
2001	0	0	0
2002	3	3	10
2003	0	0	6
2004	1	1	0
2005	1	0	4
2006	39	0	0
2007	42	6	7
2008	51	13	3
2009	108	7	7
2010	103	9	9
2011	46	3	7
2012	44	0	0
2013	2	1	0
Total	440	43	53

Note: Only January–February 2013 included.

Source: Bundeswehr 2012b, Forsvaret 2013, MoD 2013b.

The first German and Danish deaths occurred at a time when troops from both countries were deployed in relatively quiet areas of Afghanistan. As fighting began to escalate elsewhere, visitors could sometimes mistake the Germans and Danes for security guards watching over development projects; it was an image nurtured by political leaders in both countries. When the Danish Prime Minister Anders Fogh Rasmussen visited Afghanistan in May 2007, just after an IED had killed one soldier and badly wounded two others, he tied their presence in Afghanistan directly to the protection of girls' schools (Svane 2007). The Danish and German media differed sharply, however, in how they framed the deployments. In Denmark, development assistance and fighting the Taliban were both important topics; in the German media at the time, fighting was excluded. Part of the explanation was that German media coverage naturally reflected developments on the ground in German-controlled territory, where there was little action compared with the south; another factor was the political framing of ISAF as preferable to the more bellicose operations of the Anglo-Saxons.

When the first deaths were reported, they made headlines in all three of the countries analysed here. However, what the media was able to report varied greatly, because journalists were dependent on information supplied by the authorities. In Germany, few details were given apart from age and rank. As one observer remarked, it was impossible to distinguish between a soldier run over by a bus and one killed in combat (Kremp 2005). The official explanation was that the next of kin had to be informed before anybody else, and that further details might impair the security of those serving in Afghanistan. For the media, it meant there was very little to write about. In the early years there were hardly any interviews with soldiers who had served in Afghanistan, and the practice of sending off journalists to visit bases and talk to the soldiers was far less common in Germany than in the two other countries. In periods, Danish newspapers have published stories and interviews on an almost daily basis. In the German media such intensive coverage has been rare. As Denmark's dead arrive home from Afghanistan, most Danes already have an idea what the youngsters were like and what they were up against. The same can be said for the British where coverage, especially in the tabloids, remained extensive throughout the war. In German media, first-hand impressions of what was going on were comparatively rare. When Horst Köhler, the Federal President, complained in 2005 that the relationship between the population and the armed forces was one of 'friendly disinterest', the official information guidelines could be blamed (Köhler 2005). The media had simply too little to go on.

Soon a gap emerged between the rhetoric of German political leaders as they sent soldiers off to Afghanistan and the direct knowledge of those boarding the aircraft. The development aspects so praised by the government, whether the Social Democratic-led one under Schröder or the conservative coalition led by Angela Merkel that took over in 2005, could not disguise the facts on the ground. The Taliban regarded the Germans as invaders no less than the British and the Danes fighting in the south of the country (Blechschmidt 2006b).

Germany and the K-Word

When the first bodies were returned, the Danish and German authorities alike were unprepared. How should the coffins be received and by whom? What was an appropriate reception ceremony and what kind of speeches should be held? Should the press attend, or only the next of kin? Such questions had to be solved in short order.[3] The outcome in the two countries was very different. In Denmark, openness and publicity were important, and the political attendance was conspicuous. That the dead had fallen in war was not only acknowledged but emphasised. Germany chose the opposite approach and strictly adhered to it well into 2008. Repatriation there was conducted quietly; religious services were usually held without any press coverage. Although local newspapers would report on them in great detail, the national papers would not. In fact, had it not been for the detailed account of a German officer who returned in 2007 after surviving an attack that killed two of his comrades, we would know very little about how the German authorities responded to casualties up to that time (Löwenstein 2008). The coffins had created such shock, the officer discovered, that neither the government nor members of the Bundestag had settled on a procedure to receive and honour them. By then, 21 soldiers had been killed and brought home. His astonishment at the lack of ceremony was likely shared by other men and women in uniform. Banning the press from repatriation ceremonies meant minimal coverage, apart from the comments added to blogs like *Soldatenglück.de*.[4] Thus, the grievances of soldiers were read primarily by other soldiers.

To Germany's men and women in service, the lack of public attention to the dead seemed closely linked to the official policy of refusing to acknowledge what was happening on the ground in Afghanistan. In September 2008, the Parliamentary Secretary of State in the MoD labelled the fighting in the German sector as nothing more than 'skirmishes', explicitly rebutting any other term (Stuttgarter Nachrichten 2008). This contrasted with the wording used by Germany's allies. That same September, soon after the Taliban had killed a German soldier, the German Minister of Defence visited the German camp in Afghanistan and was asked by a journalist whether this was war. The minister replied that, although the soldiers were fighting terrorism, it did not mean they were at war. Standing beside the minister was US General David McKiernan, who commanded ISAF, including the German troops. He corrected the German minister in front of running cameras with a terse, 'Yes, we are conducting a war' (Hammerstein et al. 2008). The German media cited McKiernan's statement to decry the official silence over violent developments in Afghanistan. Still, the American general's remark failed to trigger any debate in the Bundestag, let alone a change in the official German line.

This did not stop returning soldiers from telling about increasingly harsh fights. The term they used was *Krieg* [war]. From late 2007, newspapers would start to address the gap between their version and the official one. Despite increasing attention, the change in official rhetoric was slow. A gradual move towards accepting the soldiers' description occurred the following year, when the press officer in the Ministry of Defence admitted that the stabilisation measures were of a particularly 'robust kind which include combat action' (*Focus* 2010). Then

3 On this I am indebted to Colonel Lennart Scheel of the Danish Defence Academy for sharing his knowledge with me.

4 Soldatenglück can best be translated as 'the soldier's good fortune'.

finally, in early April 2010, Minister of Defence Guttenberg took a significant step towards validating the soldiers' version: 'Even though not everybody likes it, when faced with what is going on in Afghanistan, at least in parts of Afghanistan, one may in everyday language – and I emphasise everyday language – talk about a war in Afghanistan' (*Focus* 2010). Qualifying 'war' in this way was needless; nobody disputed that the Bundeswehr was engaged in war. The men and women on the ground felt vindicated; and ironically, so did the government's harshest opponents in parliament, The Left party. Its members had been claiming since the departure of the first soldiers that ISAF meant war.

The belated acknowledgement of the soldiers' version of events did not represent a volte-face in the official framing of the war. The 2010 government report on the German area of responsibility emphasised, as had previous reports, that the area remained calm compared with regions where fighting was continuous (Bundesregierung 2010). This was not the only official document presented to the Bundestag about Afghanistan. The Parliamentary Commissioner for the Armed Forces' annual report provided no grounds for complacency (Wehrbeauftragter 2010). The text pointed out repeatedly that German troops lacked adequate equipment to confront growing security threats. The problem has not been unique to Germany. Complaints about insufficient equipment have been common to all Western participants in the war; often the fuss has prompted the opposition to pose a question in parliament to the responsible minister, and if the deficiency is deemed serious enough, budget allocations are made. Germany has been unique in that the Parliamentary Commissioner in periods has issued the same complaints annually, almost verbatim, to no avail (Wehrbeauftragter 2008: 13, 2009, 2010: 13, Bundestag 2011: 15, 16, 18, *Welt* 2011). His remarks have triggered little debate and apparently no fresh funds for better equipment. There has been no parallel in the Bundestag to the heated exchanges in Britain over helicopters and armoured vehicles. Although the sums earmarked for Afghanistan increased with the growth in the number of soldiers deployed, extraordinary allocations made to meet urgent needs on the ground were not made. Parliamentary Commissioner Reinhold Robbe remained alone in his requests. For a long time, this may have been due to the military leadership's reluctance to go public with demands for additional funds, fearing popular backlash (Inacker 2003). When the 2010 report was debated, the preceding 12 months, March 2009–March 2010, had been particularly bloody, with six killed and 60 wounded. A defence spokesperson for the governing party admitted that the time had come to 'face up to the facts' and finally allocate the means required to improve soldier protection (*Frankfurter Rundschau* 2010). Instead of increasing spending, allocations were cut (*Handelsblatt* 2011).

The German government's suppression of debate on the equipment issue and its refusal to permit German soldiers to assume command positions in the coalition seemed clearly designed to preserve its framing of the deployment as an assistance and reconstruction project. In 2002, information that German soldiers were involved in offensive operations with US forces emerged. The source was the US Central Command (Leersch 2002). Bundestag members and journalists alike claimed that they had been misled about the true extent and character of Germany's military engagement (Grassmann 2002, *Welt* 2002). The joint operations probably ended soon thereafter, due to the protests in Berlin. In any case the scale of the operations was insufficient to dampen criticism by allies of the German contingent's limited military effectiveness (Buck 2006, Pries and Meng 2006, *Frankfurter Rundschau* 2009, Volker 2011). In 2011, the Parliamentary Commissioner for the Armed Forces concluded that reduced budgetary spending would lead to a further deterioration of the Bundeswehr's standing (*Handelsblatt* 2011). Like his other warnings, this one had no discernible impact on the domestic debate.

When informing the Bundestag of events in Afghanistan, the government kept the focus on civilian aspects of the German deployment. It is nearly impossible to find reports mentioning the growing security danger to German relief projects. Other countries had to revise their development ambitions and there is no reason to believe that Germany fared any better (Stoddard et al. 2009). The deterioration of security for aid workers could have been used to argue for a more robust military presence. While British and Danish informational material underlined this linkage very clearly, the official German material continued to treat security and development separately (Reinhold 2009). In the government's 2006 *Afghanistan Concept*, security problems in the German-controlled area were ascribed to drug barons and smuggling, implying that these would disappear with the progress of civilian reconstruction and political normalisation (Auswärtiges Amt et al. 2008: 9). The Bundeswehr, in its own analysis, refuted as groundless the belief that northern Afghanistan would remain calm; instead it drew attention to the Taliban's exploitation of the 2005 parliamentary elections to renew its hold on members of the new elite and expand activities to all parts of the country (Schetter 2007). This contradicted the government's claim that combat in Afghanistan was confined almost exclusively to the south. The Bundeswehr concluded that its RoE must change to counter the new threats. If they did not, whatever progress Germany had made toward reconstruction would be lost (Chauvistré 2009b). Nevertheless, the RoE remained basically unaltered. As a result, the Bundeswehr was barred from participating in a number of operations, primarily dangerous ones (Minkmar 2008, Chauvistré 2009a).

German leaders who refused to acknowledge the rising threat posed by the Taliban also refused to alter the official German assessment of Afghanistan's root problems. They simply reiterated what had been said during the initial Afghanistan conference in 2001: what Afghanistan needed was reconstruction and assistance. The main obstacle to meeting those needs was war. The solution was not to end the war by defeating the Taliban, but to reduce the military presence in the country. In other words, the US-led operations and the unrelenting combat of the Danes and the British were contributing heavily to the country's problems (Wieczorek-Zeul 2003, 2008). The emphasis on military means was an Anglo-Saxon fixation, primitive and doomed to failure. The 'Continental approach' – German, that is – remained a safer path to success, with its focus on state building (Münch 2009: 6).

In Denmark and Britain, few questioned the importance of the military in providing stability to the local population so that development work could proceed. In Germany, the narrow official focus on stabilisation and reconstruction meant that the military role escaped public attention, apart from the muted homage paid to coffins returning home. This lack has been evident to close observers whenever the government has asked the Bundestag to approve an increase in the number of soldiers deployed in Afghanistan. By the end of 2011, close 4,800 German soldiers were there, making Germany into the third largest troop contributor to ISAF (Bundeswehr 2012a). One might have expected members of parliament to pose questions about the German force's mandate, the RoE and the security of individual soldiers as casualty rates escalated. But they did not. When the parliament was asked in early 2010 to consent to an increase in deployment numbers, the response adhered to a well-established pattern. The Left decried the war as an imperialist undertaking victimising the civilian population; the Greens claimed that insufficient attention was given to reconstruction; and the Social Democrats underlined the importance of sticking to the announced exit date of 2014 (Bundestag 2010b). That the Bundeswehr by that time had lost 37 soldiers, including seven in the preceding year as the security situation steadily worsened, was not mentioned by a single member.

Denmark – the Warrior Nation

Denmark is a contrasting case to Germany. In official tributes, government ministers have hailed Danish soldiers killed in Afghanistan as model young people whose sacrifice ennobles the nation. Support for the war has remained above 50 per cent almost until the end of the war. Denmark has become, in the word of one journalist, a 'warrior nation' (Lavrsen 2010). Although public opposition has remained steadfast at slightly less than a third according to most opinion polls, it has never managed to alter the national agenda or significantly influence debate. At one point more than 50,000 Danes had decorated their cars with yellow ribbon stickers inscribed, 'Support our troops' (Lavrsen 2010). In 2010 the armed forces filled 6,500 vacancies with volunteer recruits in record time. It was the first time since 1945 that the military did not have to resort to conscription (Damm 2010). One reason may be that more than 90 per cent of veterans coming back from Afghanistan have described their service as a positive experience (Forsvaret 2012).

As dead Danish soldiers have been repatriated and buried, the official ceremonial presence has remained strong and the media coverage extensive. Such attention has helped underline the dangers confronting men and women sent to Afghanistan.[5] A security scholar has concluded that 'body-bag syndrome' was no Danish malady (Jakobsen 2004). The high casualty rates have not hurt

5 Sophia Bruun was the first female member of the ISAF forces killed. She died 1 June 2010 after her vehicle hit an IED.

popular support for the war in any lasting manner. In fact, as will be argued here, Denmark's dead have played a key role in buoying support. Although officials initially described the operation as peace support, their rhetoric changed from 2006, when Danish soldiers moved to Helmand province to fight the Taliban. From then on, official speeches and publications about Denmark's mission have rarely referred to peace support as part of the troop's mandate (Reinhold 2009). Reconstruction and development projects have not been forgotten, though. The information brochure from 2010 entitled *Denmark in Afghanistan – Why, How and How Long?* emphasised the importance of assistance, then added that no aid projects could be launched without a strong military presence (Dueholm 2010). Many Danish soldiers in their writings give similarly nuanced views. Some have pressed for more combat and less camp time, for the reason that expelling the Taliban would be the best way to ensure local progress (Sterling 2008). As for their motivation to serve in Afghanistan, many cite a desire to help Afghans (Møller 2010: 318–20). Often they are eager to underline progress made in health and education, well aware that 'soft' news often gets lost among the battlefield stories. The official focus in Denmark on military prowess, courage and combat may actually have deflected attention from the development aspects of the Danish deployment and the significant progress achieved. In retrospect, the government's failure to use first-person soldier accounts to justify Denmark's deployment and strategy was undoubtedly a missed opportunity.

Nevertherless, Denmark's official communication programme would appear to be a success story in ensuring popular support despite high casualty numbers. The opposite might well have been the case. When the novelty of the country's initial deployment faded, so did the media coverage. Little seemed to be going on. Members of parliament and journalists did ask when the men and women would be coming back. In 2006, when popular uncertainty about the timeline began to grow, Minister of Defence Søren Gade stated in a newspaper article that Denmark was no quitter. It was in Afghanistan for 'the long, hard haul' (Gade and Tørnæs 2006). Until then, casualties had been low, with three in 2002 and one in 2004. Gade's article was notable in that he warned readers that Danish soldiers faced growing dangers and that more losses would ensue.

His prediction was fulfilled the following year, when seven Danes were killed. The country's most widely read tabloid responded with the front-page headline: 'How many more must die?' (*Ekstrabladet* 2007). Left-wing politicians and scattered artists and pacifists attempted to mobilise support for a withdrawal. But the response was muted. In parliament, the centre-right government knew it had the support of the largest opposition party, the Social Democrats. Danes in general had grown accustomed to the deaths, as the editor of another newspaper remarked (Brøndum 2008a). As in Germany, a friendly disinterest prevailed. The difference was that the Danish government clearly perceived the lack of popular engagement as a potential long-term threat to the war effort. Without explicit public support, political consensus could crack. A more immediate concern to the government was the effect on troop morale; some soldiers in harm's way might begin to feel unappreciated (Larsen 2008: 143–4, Birkeland 2010: 425–36). The Minister of Defence did not respond directly to the inflammatory tabloid article, but he came out with an attack on Denmark's allies for not doing enough (TV2 2007a). The only ally to respond was his German colleague, who insisted that the Bundeswehr would of course assist the Danes if they requested it (TV2 2007b).

The minister's criticism in no way signalled that Denmark would reduce its deployment or change tactics. That would have meant a complete break with the governing coalition's foreign and security policies. His comment on allied burden-sharing had been intended rather to highlight Danish combat ability and stamina. Yet pandering to national pride did not answer the legitimate underlying question of why Denmark should be at the forefront in Afghanistan. A fuller, more detailed outline of Danish policy was given in a brochure published in the summer of 2008 (UD and FD 2008). Designed for a large readership, it was available without charge at public libraries. Most of its brief, illustrated contents concerned the need for civilian reconstruction and Danish efforts to promote it. To make reconstruction possible, it said, the number of soldiers would be increased. Only then would a withdrawal be possible.

The effects of this brochure on public opinion are impossible to gauge, not least because its publication did not occur in a vacuum. Starting in 2005, a growing number of personal accounts from Afghanistan had been published. With some delay, media attention followed suit. Detailed portraits of the fallen appeared more frequently. This was done by *Berlingske Tidende* from the spring of 2009. The growth in media coverage was not limited to the newspapers supporting the deployment, such as *Jyllandsposten*. An examination of the archives of papers with a more critical stance, such as *Politiken* and *Information*, reveals the same tendency. Television broadcasters presented documentaries about the war, and soldiers who had served in it became frequent talk-show guests. The tone and contents of Danish first-person books and other public accounts have distinguished them from what has come out in Germany. Naturally, the deaths of Danish comrades receive prominent mention, but recurrent themes also include the intensity of combat and pride in being a member of a team of military professionals. Some accounts repeat the official rationale for deployment, asserting that Denmark's international standing depended on its shouldering of combat operations. Staying safely inside a token compound would not be enough (Hundevadt 2008: 51). Soldiers who previously had served tours of duty in the Balkans and Iraq have been those most likely to perceive the Afghanistan mission this way. Others see it primarily as a justified Western retaliation against the terrorist attacks on the United States: '11 September 2001 is basically why we are down here now', a Danish officer writes in his diary before concluding: 'One might say the we have come back full circle with me sitting down here in Afghanistan where Osama bin Laden once controlled his terrorists' (Lassen 2009: 37).

While the British and German media could count on men and women in uniform as a source of criticism of official policy, Danish soldiers have not played that role. An important reason has been Defence Minister Søren Gade's (2004–2010) popularity among the soldiers (Brix and Cordsen 2010, Pedersen 2010). Another, no less important reason was the early official framing of those deployed as representing a historical bridge to a more heroic Denmark. For more than 100 years, this small country's best hope had been 'to live as unnoticed as possible', in the words of Prime Minister Anders Fogh Rasmussen. The policy of self-effacement was now over (Ramussen 2003). The uniformed Danes in Afghanistan were living proof that Denmark had recovered the proud historical identity broken by its defeat to Prussia in 1864 and further by its political co-operation with German occupiers in World War II (Rasmussen 2003, 2006a). The men and women in Afghanistan resurrected the nation's honour. Many of the books, articles and interviews analysing Denmark's contribution to ISAF describe 1864 as a shameful blemish which now has been erased (*Politiken* 2006, Boas 2007, Skadhede 2007, Møller 2009: 10). A military priest who spent a half-year with Danish troops has written that their demanding mission extended the efforts of previous generations 'to resurrect Denmark after the [1864] defeat'. The long and hard fighting at Musa Qala in 2006 was presented as the definitive break with that humiliating past (Brøndum 2006).

Musa Qala was a showcase for Danish military prowess. And despite the protracted combat, no Dane was lost. The Army Operative Command announced that more than 70 Taliban soldiers had been killed (Stenstrup and Broberg 2006). This was uncomfortably similar to the US Command's use of body counts as a success indicator during the Vietnam War. Members of the parliamentary opposition objected to the announcement, claiming that the portrayal of events suggested that Denmark's deployment was primarily about warfare instead of assistance and development. Soon, the Minister of Defence sought to calm the criticism by reducing the estimate of Taliban fatalities to between 25 and 50 (*Politiken* 2006). Finally, the Army Operative Command declared that no official estimate would be published in order to prevent the impression that progress was being measured in the number of dead enemies (Brøndum 2008b). Officials said an accurate estimate was impossible anyway because the Taliban, like their Western adversaries, would do their utmost to retrieve their dead from the battlefield.

Two years later, in 2008, the Social Democrats' spokesman on defence issues set aside any reservations about publishing enemy losses, stating that 'people could stand the truth' (Svendsen 2008). Other elected officials stepped up to support him (Brøndum 2008b). Why they changed their view is unclear. One reason may have been the rising Danish casualty rate, with eight losses so far that year. Danish deaths could be offset psychologically by a high Taliban death toll. Another reason may have been the lack of clear news from the fighting between Danish soldiers and the Taliban in the city of Gereshk, which had continued for some time. Taliban losses would perhaps help justify that struggle. Despite the political pressure, however, the Army Operative Command refused to go back on its policy of keeping Taliban losses from the public.

Dead Taliban fighters would continue to cause debate. In 2010, the documentary *Armadillo* was screened all over the country to great acclaim. The film was a non-heroic portrayal of a group of young men, following their progress from training at home to combat in Helmand (Harder 2010). In a scene that became emblematic not only of the film, but also of the popular perception of what the soldiers were up against, the Taliban attacked the Danes, who retaliated by machine-gunning their attackers. In the aftermath, the Danish soldiers are shown cheering. Opponents of the war seized on the sequence to insist that the men had been turned into brutes by the war and ultimately by the government that had sent them (Jensen 2010, Værge 2010, Bredal 2011). But this attempt to rally opinion against the war failed. Soldiers who had served in Afghanistan praised the film as realistic. Commenting on the combat scene, they would say it was only natural for their comrades to celebrate when they came out of battle unharmed (Krog 2010). In the press coverage that followed, this view met with broad support. Dissenting voices were few.

Britain's Casualty Frankness

The British approach to casualties, unlike the German, has been one of openness. As in Denmark, the names and causes of death are published by the UK MoD (MoD 2013a). Names have been read out in Parliament as a sign of respect; sometimes this would occur in the middle of a debate over an unrelated topic, bringing proceedings to a temporary halt (Hansard 2009: col. 956). That all members have openly agreed on commending the dead as admirable representatives of the country has been an important signal to the rest of society. Tony Blair's statement at his final parliamentary Question Time exemplified the British way of honouring military sacrifice in Afghanistan. After naming three men killed by the Taliban in recent days, he concluded:

> Mr Speaker, since this is the last time that this, the saddest of duties, falls to me, I hope that the House will permit me to say something about our armed forces, and not just about the three individuals who have fallen in the past week. I have never come across people of such sustained dedication, courage and commitment. … But whatever view people take of my decisions, I think that there is only one view to take of them: they are the bravest and the best. (Blair 2007: col 323)

On this occasion, as on all others, MPs expressed sympathy for the families of those killed and support for the armed forces. When defence issues or Afghanistan policies have been debated, MPs occasionally refer to soldiers from their constituency who have lost their lives. Naming them is done both as an act of respect and to enhance the MPs' own standing; yet by mentioning the fallen British soldiers in debates, the MPs show they are not forgotten. In Denmark, such tributes have been made at commemorative services, but rarely in parliament. In Germany, apart from an obligatory statement from the Chancellor or the Minister of Defence that a Bundeswehr soldier has been killed, the dead have never been named in a parliamentary debate. Debate in the German Bundestag has usually revolved around the use of military force as opposed to aid and the degree to which Germany's participation has undermined its traditional emphasis on peaceful conflict resolution, whereas the British debate has been more fact-based, often reflecting current events in Afghanistan. The German and Danish parliaments have been informed, but they have not treated casualties as sufficiently grave to interrupt the debate agenda.

Some references to the fallen in the House of Commons have been meant as attacks on the government for its conduct of war. When the first British troops flew off to Afghanistan in late 2001, parliamentary support was overwhelming. The sense of imminent threat certainly dampened whatever opposition there was. Although misgivings arose over what military leaders saw as a dangerous overstretch of scarce resources, they did not surface publicly until Britain entered the war against Iraq in 2003 (House of Commons Defence Committee 3 July 2007: 32, 39–40). This proved to be a bloodier affair than Afghanistan. In 2003, 53 British soldiers were killed in Iraq while none lost their lives in Afghanistan (BBC 2009a). Over the next two years, two British soldiers would die in Afghanistan, compared with 45 in Iraq. In 2006, the losses in Afghanistan were greater, with 39 dead compared with 29 in Iraq (iCasualties 2013).

When the debates over equipment issues started around 2005, no distinction was made between the losses being sustained in the two theatres of war. A separate debate over Afghanistan, never mind its death toll, was rare. However, when accusations against the government for conducting 'war on the cheap' grew towards the end of the decade, they were often based on complaints from soldiers in Afghanistan who said they lacked helicopters and armoured vehicles (Harding and Swaine 2008, Macintyre 2008, Elliot 2009). MPs were soon claiming in the Commons that cuts to the defence budget imperilled soldier safety. The media had no problem finding officers who would support MPs attacking the Labour government for stingy defence spending. But at no point did journalists discover a crack in the political elite sufficient to undermine the official line that the soldiers on the ground had the requisite equipment. The only time journalists could print a dissenting view from within the government was Lord Malloch-Brown's farewell interview in 2009 (Riddell 2009). He declared himself to be on the side of the officers who had voiced their discontent (Riddell 2009). Had he been a cabinet member and not just Minister for Africa, his support might have made a difference. Prime Minister Gordon Brown's response to Malloch-Brown did not deviate from the official line: the troops had the resources 'to do the job' (Mason and Ashton 2009).

Funding and equipment issues have remained firmly on the British agenda (Clarke 2011). Disagreements emerged whenever defence issues were debated in Parliament, and very often during Question Time. The debates often focused on helicopters and the thickness of steel used in transport vehicles, issues that would have been too technical to attract popular interest had it not been for the media. Newspapers pursued the disputes in articles and interviews with soldiers on the ground. The impression left was that the men and women serving in Afghanistan perceived matters differently to the political leaders at home, and that the government had failed to provide the protective gear needed for survival in war. With the large casualty numbers suffered both in Iraq and Afghanistan, media attention to the human costs grew. By 2008, according to journalism professor Brian Cathcart, it had reached unprecedented heights (Cathcart 2008). Certainly, the detailed coverage of military funerals in tabloids like the *Daily Mail*, the *Sun* and the *Daily Record* was astounding. They maintained focus on the casualties more consistently than did many of the quality newspapers, which sometimes skipped coverage of a single soldier's death or funeral. Tabloids were particularly attentive to the presence of political leaders at funerals. Absence was equated with political neglect. In August 2009, *The Sun*'s front page read: 'Don't you know there's a bloody war on?' (*The Sun* 2009). Inside, the article continued with the following accusation: 'As the hearses carrying our heroes are saluted in silent Wiltshire streets, Mr Brown and his ministers are missing in action.' Other newspapers followed suit, albeit in less vitriolic terms (Drury and Wilkes 2010).

The Sun's attacks on the government for not doing enough for the troops should be seen in the context of the newspaper's sponsorship of a military awards ceremony whose nominees were chosen by the public on the basis of heroic deeds. The event was attended by members of the royal family and by government ministers. Almost five million viewers followed the 2010 ceremony on television (Denning 2010: 36). The government was well aware that its standing with the electorate would fall if it were seen as unsupportive of veterans' needs. This had been a lasting lesson from the Falklands War. In 1988, the BBC had screened *Tumbledown*, a docudrama about a soldier in the Falklands and his return home. The indifference that greeted him in Britain greatly embarrassed the incumbent government.[6] When that soldier returned to the public eye in January 2007 to accuse the government of repeating the mistaken pattern, uncomfortable parallels were drawn between then and now in the press, and probably in ministerial quarters as well (Griggs 2007).

It is impossible to gauge with any precision how many British people shared the Tumbledown soldier's perception of an unresponsive government, or whether it affected public attitudes to the Afghanistan war. Yet, as the war dragged on, disagreement emerged at the top between the government and Parliament and increasingly between government ministers and security experts. An impression took hold that decision makers were unable to agree on a long-term strategy for the British deployment. This had a negative impact on public backing, but the decline was not linear. When the war started, two out of three had supported the government's decision to send troops (Travis 2001). By 2009, 46 per cent supported the war – a lower portion, to be sure, but still six percentage points more than had been recorded

6 At a remembrance service at St. Paul's after the war, he was given a seat hidden away in the shadow. He was not invited to the Lord Mayor's victory parade (Townsend 2007).

in a poll the year before and 15 points more than in 2006 (ICM 2009). One reason for the jump was the end of UK combat operations in the unpopular Iraq War, in 2009. Public support for the two wars was generally in synch (Glover 2009). Still the rise in support for the fight in Afghanistan was puzzling because it happened at a time when casualties were up significantly and the prospects for success dimmer (Norton-Taylor 2009). The surge may have reflected a rallying-around-the-flag sentiment. That would explain why rising numbers of people would report support for British troops even as only a third of those polled believed the presence in Afghanistan was having a positive effect (Glover 2009). On this point the public seems to have been more discerning than Air Chief Marshall Sir Jock Stirrup, who claimed: 'Support for our servicemen and women is indivisible from the mission' (Stirrup 2009). Nevertheless, the surge was only temporary. In November 2011, an opinion poll showed that a majority of almost 60 per cent wanted the troops brought home immediately (Norton-Taylor et al. 2009). The result was repeated in a new poll a half-year later (Press Association 2012).

British troops, in common with their Danish and German colleagues, have been sensitive to changes in public support. In 2009, an officer in Afghanistan worried about public approval in an interview with the BBC: 'The Taliban is not going to defeat us militarily, but we want the same patience, courage and discipline that soldiers show here from the public at home' (BBC 2009b). A year later, the newly appointed Chief of Defence Staff, General David Richards, admitted that people seemed unconvinced it was a war worth fighting: 'If we lose this war it will be in the homes of this country, as people tire of it' (Dunn and Larcombe 2010). His statement came at a point when British losses had been particularly high; in 2010, 103 British were killed; the year before the number was 110. The high death toll did not lead to any change in the government's war rationale. By this time, debate was extending well beyond the question of whether the war was being conducted on the cheap. Gradually, doubts were raised over the strategy chosen, and members of Parliament asked if they had been properly informed about the adoption of a more aggressive military strategy. In 2011, the Foreign Affairs Committee in the Commons concluded that 'mission creep' had set in, with the original targets incrementally replaced by new, more dangerous ones (Foreign Affairs Committee 2011: 178). At the same time, official rhetoric was now downplaying the danger confronting British troops, especially in southern Helmand province. When the troops had been deployed there in the spring of 2006, Minister of Defence John Reid said he hoped they would be able to leave 'without firing one shot because our job is to protect the reconstruction' (BBC 2006b). An ex-Chief of the Defence Staff rebutted this assumption publicly, calling it 'cuckoo' because it presented too optimistic a picture of the menace posed by the Taliban (BBC 2006a). He believed too much political attention had been lavished on civilian projects and on the soldiers' role as providers of 'the stability which has brought five million refugees home, the advances in democracy, the economy, human rights and women's rights' (Browne 2007). This was not what the men and women had been trained to do. Likewise, in the words of Paddy Ashdown, an ex-marine and former leader of the

Liberal Democrats: 'The army's job in a war is to find and kill the enemy' (Norton-Taylor et al. 2009). But Lord Ashdown was no advocate of one-dimensional force in Afghanistan. He had in fact turned down an offer to become the UN's top envoy there because of what he perceived as insufficient emphasis on civilian assistance (Ashdown 2009). When the Conservative government led by David Cameron took office in 2010, Ashdown's views found little support. The newly appointed Conservative Defence Minister Liam Fox took the opposite view, stating that the British troops were in Afghanistan for the sake of national security and 'not for the sake of education policy in a broken thirteenth-century country' (Dummet 2010).

The disagreement over strategy was hardly a probate remedy to stop the drop in support. Military leaders tried to remain upbeat about victory prospects, even in the face of rising casualties. When the UK's death toll reached 300 in June 2010, the British deputy commander of the NATO-led force, Lt. Gen. Nick Parker, warned against allowing individual tragedies to affect the mission's plan (BBC 2010). Prime Minister Cameron struck a rather different note when he stated that it was the 'moment for the whole country to reflect' on the price paid (Collins 2010). When the 400-death milestone was reached in March 2012, the government responded to calls for a quick withdrawal by insisting the war had to continue to justify the losses. Withdrawing would mean the lives had been lost in vain, or in the words of the Defence Minister: 'We owe it to all those who sacrificed their lives to see this mission successfully concluded' (Hammond 2012). Thus, the high casualty rate itself became a key reason for continuing the war.

Conclusion

The Danish, British and German soldiers fighting the war in Afghanistan have not always seen eye to eye with their national leaders, whose own struggles were of a political nature. The most contentious gap in perception in all three countries has centred on how the war is justified – its framing. But conflict across the gap has not been automatic. Almost without exception, the soldiers who sign up for more than one tour of duty cite comradeship and team spirit as motivations (Müller 2006). But national differences emerge in how they rationalise the war. Danish and British soldiers have sometimes equated their fight against Taliban with protection of their families at home (Zøllner 2008, Aallmann 2011). Few of them explore the war's impact on Afghanistan itself, or suggest that they have fought for the sake of the Afghans. Among those who do, opinions diverge. Some point to the plight of women under the Taliban, and claim that many have been liberated from cruel oppression. Others point to medical officers treating civilians as an example of the positive aspects of their stay. When the soldiers reflect upon Afghanistan's future once ISAF leaves, there is none of the optimism expressed by political leaders at home in speeches and in official documents – views suggesting, for example, that victory is within reach. Likewise, as the coalition withdrawal date approaches and Danish and British soldiers have come home to write books and give interviews, they mostly decline to support the claims of political and military leaders that the Afghans are ready to assume responsibility for their own security. They insist they have made a positive difference locally in Afghanistan, but their prime message seems to be that the men and women they served with have shown themselves to be able soldiers. In this light they portray their deployment as a success, and the loss of their comrades as a tragic, but not unexpected part of it.

The German accounts are altogether different. They have tended to question, then reject, the official explanation of why they were in Afghanistan and what they were supposed to achieve there. The official line, that they were in Afghanistan to maintain security and assist the local population in reconstructing Afghan society, has received little support from members of the German armed forces who have gone public. Nor have the Bundeswehr veterans shown much appreciation for the RoE that effectively barred them from fighting the Taliban unless attacked. Frustration has been widespread and morale (to quote the Parliamentary Commissioner for the Armed Forces) has been low (Wehrbeauftragter 2008: 6, 2009: 4).[7] To make matters worse, for most of the war very few leading German politicians were willing to express support for the soldiers' version of events. The gap between successive German defence ministers on the one hand and their Danish colleague Søren Gade could hardly have been greater. A common attitude among Bundeswehr soldiers in Afghanistan has been that the only people who fully understand what they have gone through are their fellow soldiers (Groos 2011a and b, Koelbl 2011). The same assertion can be found in British and Danish accounts and interviews, but without the bitterness found in the German ones.

The political agenda in all three countries has filtered down to the men and women in Afghanistan with varying effects. The Danish news editor's statement in 2008 that the war was no longer front-page material could also have applied to the political debate (Brøndum 2008a). Despite efforts by fringe parties and a few lone members, Afghanistan has never caused profound political controversy. The framing of the war has changed over the years, though. In Germany, the government's view of an Afghan development project morphed into an acknowledged shooting war only as German deaths began to mount. In Britain, too, what started out as a military intervention to bring peace and stability to an oppressed people has turned into a bloody deployment that fewer and fewer citizens support. The end result has therefore been quite similar in the two big countries.

Denmark, by contrast, has fought a depoliticised war. A good indicator of this is the concluding speech made by the incumbent Prime Minister at the end of each parliamentary period where topical and often controversial issues debated by the MPs will be listed. In 2007, the PM took Afghanistan as the departure

7 The sense of frustration may account for the inscription on T-shirts worn by many Bundeswehr members: 'We may not be in the world's arsehole, but we can certainly see it from here.'

point for his speech; in the following years Afghanistan came up only towards the end of the speech, along with other non-controversial issues like climate change and improving relations with the developing world (Rasmussen 2007, Rasmussen 2011). The demise was the outcome of a process where the deployment was framed as both an important measure to boost the country's international standing – an ambition few Danes challenge – and a bridge to a militarily more potent past. Danish battlefield exploits in Afghanistan have assuaged embarrassment dating from the late Cold War years and recaptured a sense of Denmark's more glorious distant history. Such framing has been most pronounced in connection with the repatriations of dead soldiers and the unveiling of a monument to commemorate them (see chapters 3 and 4). Because this framing resonated well with the majority of the population, and because opponents of the war lacked both sound arguments and leading figures, the political elite could refrain from debating the war and its casualties.

British politicians have rarely mentioned the historical dimensions of the war. This may be understandable given that both British attempts to conquer Afghanistan, in 1842 and 1878–80, ended in defeat.[8] However, within the army, the memory of these old wars has lived on. In 2006 a group of British soldiers set out to place wreaths at the scene of a battle fought in the 1842 war (Etherington 2006). Oddly enough, in a Danish brochure published by the armed forces, a reference was made to Winston Churchill's time as an officer in the Pashtun region, then the edge of British India. To show how little has changed since 1897 and why the war would be protracted, the Danish brochure quoted at length Churchill's description of the Pashtuns (he called them Pathans) as continually in conflict with each other and with neighbouring tribes, never to be trusted, and unwilling to enter any peace agreement (Reinhold 2009: ch. 3).

In Germany, only the leftist opposition tried to draw parallels between the Afghan deployment and past wars. The criticism varied little in content, characterising the mission as either an effort by the government to ingratiate itself with the US president or a reinvention of German militarism. The government made no attempt to counter such views in the Bundestag. The dead were not mentioned by name and were rarely invoked in parliamentary debates. The parliamentary consensus gained by Chancellor Schröder after his confidence vote in 2001 had clear limits, and the harsh realities of the war went unmentioned for as long as possible.

Whenever the dead have been transported home for burial, the relationship between soldiers, politicians and the general public has shifted. The change is primarily one of attention. The war in Afghanistan may have fallen off the daily political agenda, but popular attention resurges when soldiers killed in action are repatriated and buried.

---

8 In Britain, Kipling wrote the following lines about the 10th Hussars drowning in the Kabul River in 1878: 'Gawd 'elp 'em if they blunder, for their boots'll pull 'em under, By the ford o' Kabul river in the dark.'

Chapter 3
Homecoming for Heroes?

The soldiers killed in Afghanistan have all been brought home; no one has been left behind. Even bodies dismembered by explosions have been retrieved and repatriated. In previous wars where the number of dead was much greater, returning every single victim was impossible. Bodies were buried close to where they fell. Rupert Brooke's *The Soldier* refers to this practice:

> That there's some corner of a foreign field
> That is for ever England. There shall be
> In that rich earth a richer dust concealed

Only when wars ended would the remains of the fallen be gathered and transferred to a war cemetery, usually nearby. The long rows of crosses in Flanders are perhaps the most harrowing memorial to a war whose great number of dead made it impossible to return every fallen soldier. For years after, grieving relatives and friends would travel abroad to war cemeteries in what became known as battlefield tourism (Lloyd 1998).

Failure to retrieve the dead distresses family and friends who are left with no remains to mourn, but it also leaves a collective scar for the country. Efforts by United States to bring its dead home from Western Europe after 1945 suggest how painful this can be, as does the still unfinished work of recovering the remains of Americans killed in the Korean and Vietnam wars. In Europe, likewise, German and British war graves commissions still search for the remains of dead soldiers so they can be transferred to war cemeteries whose burial places can be tended and visited. The British commission and a similar American one play an integral part in official policies to honour the fallen, while Germany's commission has adopted a more didactic approach, as expressed in its motto '*Versöhrung über den Gräbern*' [reconciliation over the graves] (Wernstedt 2009).

Retrieving the dead from the battlefield to give them a final resting place seems to originate in a collective need that nobody disputes. There is nothing new about it. Sophocles' Antigone risked her life defying a decree by King Creon that her brother's body 'should not be buried in a tomb, nor mourned' (Sophocles 2003: 5). Antigone felt compelled to bury her brother by a debt to the dead: 'I will have to please those below longer than those here.' The modern accounts that I have examined echo her sentiment, though less poetically. All soldiers regard getting the bodies back to base as a moral obligation to their dead comrades (Kiley 2009: 59–60). Sometimes the retrieval must take place under heavy fire:

> … he was just incredibly heavy. I could not get him off the ground, not properly, not enough to get him onto my shoulder. Panic set in. I thought: Fuck! There's a lot of gunfire going on and I can't move this bloke. I started to do what I should have done immediately and dragged him by his webbing. (Rigg 2010: 169–70)

One British officer in Afghanistan was given a medal for returning to the battle zone five times to recover casualties; another officer received his award posthumously for attempting a similar feat only to be killed by the Taliban (McNab 2009: 348–9). Nor does a soldier's obligation to dead comrades end with his getting their remains safely back to base. That is clear from the importance attributed to attending the base ramp ceremony, after which the coffins are flown back home. There, veterans and men from the same unit will serve as pallbearers during the funeral.

The first to notice and respond to the reception given to the dead are other soldiers. Anything that can be interpreted as denigrating the memory of a fellow soldier is likely to trigger a knee-jerk reaction. When a German officer publicly expressed disappointment over the lack of official attendance at funerals, a Ministry of Defence official dismissed it as a case of 'getting too close to the microphone' (ARD 2010). Yet complaints like the officer's are a reaction originating in the importance of honouring dead comrades-in-arms. On internet forums, including private blogs and regimental websites, coffin receptions and funerals are described in great detail. In Britain and Denmark, high-level presence at such events has been interpreted as official recognition of the dangers confronting the armed forces in Afghanistan. A eulogy by a senior officer, sometimes the fallen soldier's commander, is generally printed in full or quoted. Too few words would be construed as an attempt to downplay the importance of the sacrifice. Those serving in Afghanistan would immediately decry the withholding of recognition, both from the dead soldier and from their own efforts.

In the soldiers' own accounts of their experiences in the field, the death of comrades plays a major part. This is not only because the killing of a friend or acquaintance is horrific, but because reverence for the dead is an important part of a soldier's professional identity, as many of these accounts suggest. Scholars rarely address this aspect of the casualty issue, a strange omission given how far soldiers have shown they are willing to go to retrieve bodies from the battlefield, and how ritually they later attend annual memorial commemorations. The few who have described this work, have first-hand-experience of what it involves (Dowling 1989, Sledge 2005: 161–71, Kemp and Hughes 2009: 15).

Awareness of death – or, more prosaically, the knowledge that you or the soldier standing next to you may be hit by a bullet, and that the road you travel may be littered with IEDs – is shared by all men and women in uniform, irrespective of nationality: 'You try and explain what it was like where you were but people have not got a clue', a British soldier explained (Snow 2012). The barrier to sharing cannot be removed entirely, but the way a country receives its dead makes a difference to the soldiers who survive. Apparent subtleties in official communication can speak volumes. One must look beyond linguistics to explain why for years German soldiers returning in coffins were described simply as 'dead' or 'killed', while British and Danish returnees were more likely 'fallen'. The difference has profound implications for the relationship between armed forces and society. Even if soldiers feel secure about the value of their service abroad, official language can affect whether they feel that political leaders and the rest of society share their perceptions.

The reception of bodies by a home nation is more than a matter of acknowledging the loss of life; it is also an important part of the official framing of the war. The level of official participation and the degree of public and media access all contribute to the sense of national recognition that a sacrifice has been made on behalf of society. Whether the population at large approves is another matter. Even a high degree of public participation at a reception ceremony or other commemorative event does not necessarily equate with support for the war. That of course has not stopped governments from publicly interpreting high turnout in just that manner.

The importance becomes clear when we compare how Britain has received its dead with how the two other countries have received theirs. Early on, silent crowds filled the pavements of Royal Wootton Bassett as hearses passed by with the coffins of fallen British soldiers. In Germany, the authorities did not know how to respond when the first dead were transported home, except to keep a lid on all details and effectively bar public attendance. In Denmark, reception ceremonies at the airport would be broadcast on the evening news with particular attention to the high-ranking officials present. The popular response in Britain, and Denmark, has bolstered government claims that the soldiers were sent to Afghanistan on behalf of the people. In Germany, no similar public display was initially contemplated, possibly from fear that it might trigger anti-war demonstrations. Moreover, demonstrations or not, any attention directed to the dead would undermine the official rationale of the deployment as assistance rather than war.

How the early homecomings were arranged certainly had an effect on popular perceptions of the war. In Britain and Denmark, the dead were officially lauded as heroes in articles and political statements; in Germany, the government announced its regret in communiqués and in parliament, using a few set phrases. Only gradually did greater openness come to Germany. But words are only part of the message. Much care has also gone into visuals like official attendance, bowed heads, flags and hymn singing. Are the dead to be received with full media attention or with as little as possible? Is the route of the funeral cortege from the airbase to be publicised in advance? Who will go to the church, carry the coffin, give the eulogy? Official procedures influence not only how casualties are perceived, but if they are noticed at all. When coffins have been brought home from Afghanistan without any journalists or photographers present, editors have rarely allotted more than a few lines; on the evening news, a presenter would likely have mentioned that one or more dead soldiers had been brought home, then passed quickly to another story. The opposite situation, with full official attendance from the head of state down to the local mayor, would have conveyed a very different message to the country.

With their credibility at stake, the governments had little margin for error; and to complicate matters further, they shifted in the course of the war. Homecoming ceremonies are a balancing act between the need to pay homage to soldiers killed in the line of duty and the perceived impact their coffins may have on domestic opinion. This chapter is about how the British, German and Danish governments tried to strike that balance, and the messages they sent out in the process.

Death at Close Range

Soldier accounts differ in perspective from the official version of events. At times they clash – as they did in August 2008, when a German soldier shot and killed an Afghan woman and two children who had failed to stop at a checkpoint. Immediately and publicly, the Minister of Defence expressed regret and offered to pay compensation (*Tagesspiegel* 2008). Shortly before the checkpoint incident, a German soldier had been killed by the Taliban. The official response was a routine, brief communiqué in which the Minister of Defence focused on the importance of the deployment to Afghan development (Bacia 2008). There was no acknowledgement of the dangers confronting German soldiers or mentioning of the 28 Bundeswehr soldiers that had lost their lives by then. The statement was perceived by Bundeswehr soldiers in Afghanistan as well as the German Federal Armed Forces Association as tantamount to official lack of support (Graw 2008). To them, the two fatal acts in context made it clear the Bundeswehr was in the middle of a war, one in which civilians might at times be killed. In their blogs, members of the armed forces attacked the official version for not sufficiently recognising the chaotic circumstances that could lead to civilian deaths (Dirk 2008). What was missing, they felt, was an official expression of sorrow and regret over the dead soldier and not only the Afghan civilian.

The difference between the two parties, soldiers and politicians, is not just one of perspective but also one of proximity to death. For the men and women serving in Afghanistan, death was a constant presence. Clearly, it cannot exert the same influence on politicians' mindset. Few of them are affected by the news of another casualty apart from a few minutes of respectful silence in parliament when the name is read out. If a fallen soldier comes from one's own constituency, attendance at the funeral may be mandatory. In the population at large, the repatriation ceremonies and funerals of lost soldiers have functioned as a reminder of the costs of the deployment – a reminder that has threatened whatever support the government enjoyed when it decided to send the soldiers in 2001. It was for this reason that British Minister of Defence Bob Ainsworth warned in 2009 against letting the mounting casualties deflect attention from all that was at stake in Afghanistan (Ainsworth 2009). The problem for the government was that what was at stake, presumably the objectives defined when the soldiers were sent off in 2001 fell out of focus as the number of deaths increased and victory prospects dimmed.

Ainsworth's statement, by no means untypical, is interesting because it reveals a perception gap between politicians at home and the soldiers in Afghanistan. At home, in the official version of the war, casualties are regrettable but rarely defining events. From a soldier's perspective, they are central. In books written by soldiers, the death of a comrade is almost always a turning point. The phrases are usually brief and factual: the vehicle was burnt out and he didn't make it; he was hit in the head and killed instantly; we tried to save him but he bled too much. A few have nevertheless managed to give readers a vivid impression. A British soldier writes:

> It was the worst thing I had seen in my life … . He was lying back and I was shouting at him. "Jacko, look at me, look at me, don't fucking go, stay with." We were trying to keep him stable, trying to keep him awake … but he was quiet, really quiet. He had gone white and he was looking up. (Bishop 2008: 141)

The shock of finding somebody shot or blown to pieces in a battle situation 'is to put those horrible feelings in a box at the back of your mind and close the lid. You can deal with all the horror later' (Danziger 2010: 233). It is then left to the reader to imagine just what the soldier saw and the despair he felt trying vainly to save a friend. The poet Walt Whitman wrote after the end of the American Civil War that 'the real war will never get into books' (Whitman 1892: 101). Conveying the shock and suffering seemed impossible to him. Whitman did not participate in the fighting, but nursed the wounded. If even he found describing suffering and death an insurmountable task, it is easy to imagine the obstacles facing men with lesser literary talent as they try to write down what they have been through.

When going though the descriptions of combat and loss, the question that arises is to what extent they can be trusted. Robert Graves who wrote one of the most widely read books on how World War I was experienced in the trenches, warned against taking any of the descriptions of the war as objective truths:

> The memoirs of a man who went through some of the worst experiences of trench warfare are not truthful if they do not contain a high proportion of falsities. High-explosive barrages will make a temporary liar or visionary of anyone; the old trench-mind is at work in all over-estimation of casualties, "unnecessary" dwelling on horrors, mixing of dates and confusion between trench rumours and scenes actually witnessed. (Graves 1930: 32–3)

The experience of combat is certainly deeply subjective. However, there are sufficient similarities between them to accept them as valid descriptions both of what actually happens during an attack and how the loss of a fellow-soldier is experienced up-close. Moreover, despite Graves's warning, there are no attempts to over-estimate casualties; nor any 'dwelling on horrors'. This applies in particular to the description of death experienced at close range.

In books and interviews alike we hear about the attack that killed someone and often very nearly the author as well; about the report back to camp that one man has been badly hurt or is already dead; and about the ceremony in the camp before the coffin was sent back home. Such accounts provide insight into how the writers perceive their own role in Afghanistan and how they believe society should regard it. Apart from a few overtly 'macho' tales (which form a distinct minority, often written by journalists or others not directly involved in combat operations), most are devoid of hero-worship or attempts to idealise the war.[1] The operations described of course reflect the writer's wish to present himself and his comrades in the best possible light, as upright, competent soldiers. But there is little hyperbole. The attention to detail varies among authors, but almost without exception the main emphasis is on the emotional impact of seeing someone killed. The description of Jacko's death given above is a typical example.

Rarely do the writers make any reference to their own mortality. Only in response to direct interview questions do they admit fearing injury or death. It is to be expected that such frank sentiments will loom larger in future books by soldiers. After previous wars we have seen not only that memories change with time, but that they expand to acknowledge feelings of insecurity and fear (Merson 2008, Tran 2010). In the accounts available now, what dominates is the writers' concern for members of their own units. When the soldiers of Afghanistan write memoirs in 20 or 30 years' time, what they remember will have changed. Assuming that the passage of time affects these soldiers as it has their predecessors, future books will contain less combat and more sorrow. Opening 'the lid' becomes easier. Despite the transformation of memory over time, however, death itself is likely to retain its central place in storytelling. The accounts we have today, are still raw.

Accounts coming out of the war have tended to describe the soldiers who lose their lives as good comrades, both supportive and reliable. Rank is rarely given much importance, though sometimes a fallen officer is portrayed as an exceptional leader whom the men would follow no matter what (Kristensen 2007). The pervading sentiment is one of mutual responsibility: that the soldier beside you can be relied upon: 'You've had people out here storming Taliban positions to save wounded comrades. And the reason they do it is because of the extreme feeling of friendship, camaraderie, brotherhood, whatever you want to call it, for somebody who has a similar life to you' (Barnwell 2010: 384). The team is always emphasised, with the loss of a soldier representing a gap that cannot be filled (*DefenceFocus* 2011: 5). Because of esprit de corps, the death of a fellow unit member is felt much more strongly than that of some other soldier:

> Our dead were the first from 3 Para BG and I think my initial contact report shocked the battalion Ops Room. It is dreadful but one had this awful relief that the dead weren't A Company but Royal Signals attached, and of course I didn't really know them in the same way as my own. (McNab 2009: 55)

1 Machismo and glorification of war play permeate Ramussen 2006, Møller 2009.

Loyalty to one's fellows is sometimes mentioned as a key motivator for going to Afghanistan: 'The whole reason I went to Afghanistan was to justify the soldiers who went before me. Why should I sit with my comfy slippers on not having done my bit' (Snow 2012: interview 13). The style of this quotation is quite common; the high-flown phrases of politicians are rarely found in the books and blogs.

The neutral factual tone is retained when soldiers refer to their own funerals and what should be done with their belongings after death. Before being sent off to Afghanistan, for example, they must fill in a form asking about burial preferences and other special wishes, but the replies are rarely long: 'I want to be buried next to my father. All my belongings should be divided among the boys, the money to be put in the jar. The inscription on my tombstone must be "fallen in combat"' (Stougaard et al. 2010: 432–3). Common to all is the wish for a religious ceremony, the national flag covering the coffin, and the men with whom they served as pallbearers.

In some books, the authors include photos of ramp ceremonies in Afghanistan and sometimes of funerals back home. Other photographic material, used in books written by some Danish and British soldiers, is more surprising. In between standard images of dusty deserts, poor villages and poppy fields are photos of men who are either wounded or shocked after having emerged from battle and photos of men smiling, resting or sleeping, with captions underneath noting that they were killed a short time later. The fact that many of those pictured are now dead can easily escape the reader unless care is taken to read the captions. The dead continue to be embraced as members of a team, and are remembered that way in the body text as well. The news media's approach is very different when displaying photos of the dead or those near death. The intent in that case is usually to have an emotional impact on viewers, often with front-page treatment like that accorded the shocking photos of office workers jumping to their deaths from the Twin Towers on 11 September 2001. The photos in the British and Danish books about Afghanistan have a different purpose. They are sad, possibly uncomfortable reminders of what the war entailed. It is well-known that soldiers have taken photos of battle scenes, including their dead comrades in previous wars, but these were intended as private mementos and only made publicly available long after the war ended (Koureas 2007: 158–64). The Afghanistan war is different in that this time the men have published photos of their dead comrades before the military operations were over. Since the reading public in both Britain and Denmark have been fully aware of the extent and ferocity of combat in Afghanistan through the media coverage of funerals, the shock effect of book photos is minimal if not absent. The lack of such coverage in Germany may explain why there are no German photos of dead or wounded Bundeswehr soldiers in any of the books published.

Death – Ours and Theirs

The dead are not equal. When politicians and journalists refer to them, they discriminate. Most important are one's own boys. The killing of a foreign soldier may receive public mention, but rarely with any of the detail or background provided when a countryman dies. An exception occurs if the foreigner dies in a fight involving soldiers from one's own country; then at least the circumstances of death will be mentioned. Whatever the facts on the ground, units from home always seem to play the key fighting role, with allied soldiers as anonymous contributors. The same differentiation is evident in the soldiers' own accounts. Losses suffered by coalition partners are sometimes only fleetingly mentioned. More attention is paid to a foreign soldier known personally to the author. If it's someone he had co-operated with, or who had been of assistance, then the dead was a comrade, one of his own (Beattie 2009: 256).

The multinational perspective on service deaths is missing from media coverage and soldier accounts despite the fact that many nations contribute to ISAF. News consumers at home all too easily may get the impression that their country is fighting single-handedly. This biased view of events is explicable, given the practice of embedding journalists with their own nation's troops. The battle of Musa Qala, which came to play a major role in the Danish government's framing of the war as one of heroism and national honour, was almost always portrayed in the Danish media as if it had been a purely Danish undertaking. In fact, Musa Qala had been fought over before the Danes arrived, resulting in the death of eight British soldiers, and after the siege was broken the British replaced the Danes (Rytter 2010). By the same token, the Danish role never made it into the headlines in Britain. Even in detailed accounts of the British fighting in Musa Qala, the role played by the Danes is left unmentioned, for example (Kemp and Hughes 2009). There is nothing new in this; war reporting has always tended to inflate the role played by a nation's own soldiers (Rivenburgh 2000). It's never the West that is winning in Afghanistan; it is Britain, Denmark or Germany.

The national frame is strongly present in photos illustrating most of the books by Afghanistan war participants. Men crawling in a muddy field look pretty much the same wherever they come from, but when they are photographed at the ramp ceremony taking leave of a dead soldier, national symbols dominate the frame. The coffin is draped in the national flag with a guard of honour waiting to load the coffin onto the airplane. When the funeral at home is photographed later, flags again will predominate, often supplemented by regimental banners bearing the names of famous historical battles. Their display at the funeral signals that the lost soldier is part of a tradition.

Instead of displaying explicit images of dead or wounded soldiers, German book authors tend to resort to proxies such as a cross, a coffin or a nondescript photo from an Afghan village, road or desert with the caption that this has been the scene of fighting. Such indirectness reflects the reluctance to address casualties so evident in German political debates and speeches. This will be discussed in subsequent chapters; suffice it here to mention that the men and women of Germany who served in Afghanistan have published noticeably fewer books than their counterparts from Britain and Denmark. German books and articles analysing and criticising the war are numerous, but even in those, death rarely appears except in a passing mention of casualty rates. No German newspaper has run anything comparable to the illustrated catalogues of the dead that have appeared in Denmark and Britain (*Politiken* 2011a, *Telegraph* 2001–2012). The Danish book *De faldne* [The Fallen] gives a detailed and lengthy description of all those killed, how they lost their lives, what they were like out of uniform and how their families coped with the loss (Stougaard et al. 2010). In Britain, an Internet site that was launched in 2008 presented the country's fallen soldiers and let visitors post their own recollections of the dead; the site has been regularly updated since 2008. British and Danish newspapers update their lists with each new casualty reported, whereas those published by the German *Frankfurter Rundschau* and *Spiegel Online* have not been updated since 2009 and 2011 respectively (Rundschau 2009). Undoubtedly, the British and Danish publics have much easier access to information than the German.

A casualty group rarely mentioned is that of Afghan civilians. When they do appear in the soldiers' books, it's usually as the victims of Taliban attacks, proving Taliban cruelty (Harnden 2011: 110–11). As a rule, Afghan civilians remain anonymous. Civilians killed because they worked for change and progress, such as female teachers or midwifes, are occasionally named or photographed (Hundevadt 2008: 234). These Afghans help justify the mission in the eyes of the soldiers, as do Afghans who have dared defy the Taliban while escaping harm to date. In many written accounts, the soldiers admire civilian courage and cite the protection of brave Afghan progressives as a reason for the West's military presence (Morgan 2010: 186). On the few occasions when the British and Danish prime ministers have referred to civilian casualties and their own troops in the same speech, the dead Afghans have come across as victims of Taliban terror and the Danish/British soldiers as agents to prevent such terror from happening (Svane 2007).

Western aid workers compose a numerically small group of casualties. The number of Afghans teachers or health workers killed by the Taliban has been far greater. Nevertheless, the national frame again brings considerably more attention to the lost Westerners, with special focus on the way they were killed. 'Bodies riddled with bullet holes were found in a desolate mountain area. One of them was a German woman …' opens a lengthy German description of the abduction and killing of a group of Christian aid workers (Rundschau 2010). It is far more detailed than most descriptions of Bundeswehr casualties, it may be added.

Seldom are the Afghan civilians killed by western forces mentioned in the soldiers' own accounts. Civilians wounded by mistake as well as buildings and vehicles damaged have received passing comments, always with the added remark that the losses were compensated and wounds treated. Media coverage is not much different. Civilians tend to make the headlines only if the circumstances surrounding their deaths are particularly horrific, as in the bombing of a wedding party mistaken for a Taliban meeting, or the shooting of a family in a car that failed to stop at a roadblock. A civilian killed by a stray bullet during a skirmish rarely gets more than a line of notice, and sometimes not even that.

Civilian casualties in the Iraq War were significantly more visible. The Internet site 'Iraq Body Count' provided running civilian death toll updates, and became a widely used point of reference. Attempts were made to create something similar for Afghanistan, most notably at iCasualties.org, but that site has never been relied on the same extent for civilian casualties. A valid explanation is the difference in journalistic working conditions between the two countries. Journalists moved more freely in Iraq than in Afghanistan, and Iraqi hospitals could be trusted to provide reliable death counts. In Afghanistan, most of the countryside has been off-limits to journalists travelling alone, and the numbers provided by locals are often impossible to verify. Blaming the lack of verifiable casualty numbers entirely on official manipulation, as the political scientist Judith Butler has done, is therefore overtly conspiratorial (Butler 2010: 39, Jensen 2010). That is not to say authorities have never tried to minimise the number of civilian victims of Western forces. The German Kunduz bombing is an example. Another was the official foot dragging when the Danish parliamentary opposition tried to launch an investigation into the number of civilian casualties after the battle of Musa Qala in 2006 (Nielsen 2006, Forsvaret 2006: 7082).

Notwithstanding the journalistic difficulty of identifying Afghan civilians, there is no overlooking the fact that civilian death and suffering can undermine the concerted framing of a nation's soldiers as brave, determined and caring. The image of Western soldiers as protectors and helpers was tarnished on one occasion in the German Bundestag, when the names of civilian victims of Western bombings were held up by Left MPs (*Spiegel* 2010). Political and military leaders were fully aware of the political danger. In 2008, Human Rights Watch concluded: 'US and NATO forces in Afghanistan have had a very good record of minimizing harm to civilians' (2008: 29). The following year, US General Stanley McChrystal issued strict directives forbidding air strikes except in the direst circumstances (Schachtman 2010). In the following two months, according to US military estimates, the number of civilian casualties fell sharply (King 2009). Such efforts to limit the number of civilians killed have marked a momentous change from the Vietnam War, when innocent victims were often included under the heading of communists killed, and body counts were presented as synonymous with victory (Bohannon 2011).

In exceptional cases, some civilian victims are accorded almost iconic status as the face of war. In Iraq, Ali Abbas, a little boy who had both arms amputated after an American rocket wiped out his family, made it onto most front pages. Decades before, the terrified Khim Phuc running with other children after being scorched with napalm became the human face of the Vietnam War. The only civilian Afghan victim whose portrait has been reprinted so many times that it may achieve almost the same status is a victim of Taliban, not coalition, brutality. The photograph of Bibi Aisha, a girl who had her nose cut off for insubordination, was put on the front of *Time Magazine* (Bieber 2010). The cover line – 'What happens if we leave Afghanistan?' – was answered by the girl's mutilated face. The political message was clear: the Western engagement is an instrument to protect the weak, and as a consequence any reduction in the war effort will only harm them. When the girl was later photographed with a prosthetic nose, her restored visage showed the positive effects of the Western engagement (Bates 2010).

Dead Taliban

The sight of dead Taliban fighters has prompted a variety of reactions. The scenes of jubilant Danish soldiers captured on film are typical of the reactions by men who have survived combat unscathed (Roed 2007, TV2 2011).[2] The emotions displayed were roughly consistent with those of British soldiers, as described by their colonel, after the soldiers discovered the remains of a Taliban group they had defeated after protracted fighting:

> … the spirits of the soldiers were lifted by the sight of the Taliban dead. These were the first dead bodies many of his soldiers had ever seen. This was the physical embodiment of their success. They could see the results of their work that day; these were the enemy who had been trying to kill them just hours earlier. (Kemp and Hughes 2009: 88)

The personal experience of killing is rarely mentioned by soldiers with more than a few words, and those chosen are typically devoid of emotion, with 'taken out' and 'neutralised' as the preferred terms (Møller 2010: 138–97). A journalist embedded with British troops observed: 'As for the act of killing the enemy, it is usually described very matter-of-factly, as part of the job …' (Allen 2010: 45). Among coalition soldiers there have been subtle differences in attitude toward dead Taliban fighters. If one's own unit has done the killing, it is common to express satisfaction over a mission accomplished. The matter-of-fact tone breaks down most notably when an action is undertaken to avenge the death of a comrade. A Danish officer reports that after losing a highly respected fellow officer he started to feel joy every time a Taliban fighter was killed. He 'did not care for the feeling' but considered it inevitable: 'At the bottom of, it is all about the man standing next to you. If they kill my comrade, they will be hit over the head. I am not proud of this, but this is just the way it is' (Hundevadt 2008: 175).

When he was new to Afghanistan, a Danish officer writes, he reacted negatively to American soldiers who shouted, 'Take that, Motherfuckers!' when an attacking Taliban unit was subjected to air bombardment. After having been exposed to battle, however, the Dane began responding the same way (Johannesen 2008: 213). Reflecting on his own transformation, he writes that he and other soldiers did not cheer the killing of 'another human being, a father, a husband or a son. That it's not what it is about. The cheering must be seen as a defence mechanism, it means we can continue to expose ourselves to danger'

2 See p. 46.

(Johannesen 2008). He adds that racist comments made in similar circumstances should be regarded in the same way.

Of all the accounts I have gone though, his is the only one to address racism. Undoubtedly, self-censorship plays a role here. In the books, sweeping statements about the Afghans are frequent. When asked to compare their experiences in Iraq and Afghanistan, a group of Danish soldiers all expressed strong contempt for the Iraqis but considerable understanding for the predicament of local Afghans, who risked death for disobeying the Taliban (Birkeland 2010: 400–412). Again, a gap is evident between political perceptions and the view from the ground. The political rhetoric employed in Western capitals after the September 11 attacks often dehumanised the Taliban; soldiers who have been in the field paint a picture that is notably more complex (Steuter and Wills 2010, 2011). To many of them, the Taliban's hold on the population was the result of poverty and a lack of protection by the Afghan government. The soldiers' role in Afghanistan was to offer that protection, thus preventing the Taliban from taking over areas and recruiting new members. Nevertheless, despite the sympathy displayed by soldiers for local inhabitants press-ganged into serving the Taliban, their books treat Taliban forces as a uniformly dangerous enemy to be driven out or killed.

The soldiers employ different markers for mission success. In their writings, all include accomplishments like opening a school or ensuring a village water supply. But in the books and accounts that focus on combat (and there are few that do not fall into this category), the chief indicator of success is reducing the Taliban's ability to conduct operations. Pride is expressed in the marksmanship needed to 'take out' a sniper at long distance or killing Taliban fighters planting a bomb alongside a road used by fellow troops. In descriptions of the Taliban as opponents in the field, emotions leave way for professional comments on enemy tactics: 'They are very accomplished at moving into firing positions using good cover and it is almost impossible to identify the firing points' (Harnden 2011: 155). How Taliban fighters, dead or alive, are described does not vary much. They are the enemy, and as long as they are not being killed or captured, they are relentless opponents. On this point authors from the three countries betray no national differences or any softening of view with the passage of time. Books and interviews from 2012 are no less severe in their judgement of the Taliban and how they should be dealt with than those from 2002.

With regard to the Taliban's character, the perception gap between the men and women on the ground and the politicians back home has widened as the years have passed. At the beginning of the war, there was no discord. President Bush denied combatant status to the Taliban; they were not to be regarded as soldiers but as terrorists. When Tony Blair justified his government's decision to join ranks with the United States, he portrayed the Taliban as a threat to Western values: '… our beliefs are the opposite to theirs. We believe in reason, democracy and tolerance. These beliefs are the foundation of our civilised world. They are enduring, they have served us well, and as history has shown, we have been prepared to fight when necessary, to defend them' (Blair 2001: col. 606). Blair's us-versus-them distinction was echoed by Danish and German government members. However, as the war has begun to approach an end, the Taliban has been transformed in the official view from the West's antithesis into a potential partner. As the prospects of a clear military victory fade and the need for a negotiated settlement increases, the framing process has been revised. In late 2011, US Vice President Joseph Biden stated that 'the Taliban per se is not our enemy' (Gelb 2011). European politicians also started to underline the need to include the Taliban in negotiations conducted with the Afghan government (Bloxham 2011, Rayment 2012).

Soldiers on the ground have not changed their views of the Taliban. To them it remains a dangerous enemy whose members are to be killed or driven out. This view does not keep Western soldiers from sometimes revealing a surprising degree of empathy. A British soldier who survived a suicide attack reveals just that in this description:

> There were the worst sights I have seen in my life. There's not a lot you can do when you're picking up the broken limbs and body parts of ten people. I saw the hand, I think, of the bomber. And it was an awful thing to think this had just been a person who once had loved and laughed just like any other human being, and now was a mess on the floor. (Kiley 2009: xiii)

The empathy may reflect an understanding shown by many writers of the dire conditions in the countryside; it may also be due to the arbitrariness of death noted by nearly all who have come home from the war and written about it. Earlier generations of warriors have made the same observation. Wilfred Owen, in one of his best known poems from World War I, *Strange Meeting*, expressed it thus:

> I am the enemy you killed, my friend.
> I knew you in this dark: for so you frowned
> Yesterday through me as you jabbed and killed.
> I parried; but my hands were loath and cold. Let us sleep now …

Retrieving and Repatriating the Dead

Retrieving the dead from the battlefield is a task rarely described by those who have done it. The gruesome details serve as an effective and long-lasting barrier to expression. Only with the passing of time do a few come forward with their memories. One who was part of a US army team ordered to collect the remains of allied soldiers in Northern France in 1944 has given an account not only of the task but of the impact it had one those ordered to execute it: 'Over to the left I could hear someone throwing up, and it was not from seasickness. We were over that now. This was revulsion. All the training in the world could not stand up to the real thing' (Dowling 1989: 38). His experiences are unlikely to be any different from those who have collected the remains of soldiers killed in Afghanistan and brought them back to base for the final journey home (Harnden 2011: 292). Another aspect unlikely to have changed is the sentiment that the work is important: 'Under it all, there was an odd sense of pride building up' (Dowling 1989).

The devotion to fellow soldiers killed or maimed sometimes results in actions that defy logic, at least when seen from a civilian perspective. The best example is an episode described and photographed by a British officer, showing the men of his unit burning the foot and leg of a soldier almost killed by an IED:

> We stood all around the rubbish fire during the cremation, my hope being that however perverse or shocking this may be we would do it together. It was our farewell. Later we took turns to individually stand guards over the fire. It was a refreshing reflective solitude that helped us all. (Allan 2009: 100)

As the quotation reveals, this rite was performed out of a sense of obligation; not even the foot was to be left in the open. And as the author adds, the cremation was important to all who participated, generating something like the 'odd sense of pride' felt by the army team in 1944.

After having been brought back to the camp, the body was placed in a coffin ready for transport home. Accompanying it would be a letter from the commanding officer to the family in which the officer attempts to answer some of the questions those at home are bound to have, the most important being whether the their loved one suffered a painful death. 'I find these letters incredibly hard', one British officer explained. 'I finish off by saying how much we appreciate his sacrifice, that we appreciate their sacrifice as a family, and that we are a regiment that remains on hand to support them, we're a family regiment' (Danziger 2010: 221).

Attendance has been the universal norm whenever a coffin leaves camp (Chapman 2010: 300, *Frankfurter Rundschau* 2010). No distinctions are made between the dead of one's own unit or country and the others sharing the camp. Everybody who could possibly attend the ramp ceremony does so, even when the send-off is conducted in the middle of the night. What seems to be a set pattern followed by all is described by one British soldier as follows:

> … either the day of a death or the following day we will all down tools and congregate on the helipad. The padre will lead a short service of remembrance and then someone will read out a bit of eulogy for the soldier or the soldiers who have been killed. (Danziger 2010: 218)

The bugle major who has played *Last Post* at the beginning of the ceremony, will end by sounding the Advance: '… and that's it, the service is finished, and we're off again'.

The speeches made when coffins are sent from Afghanistan and when they are received at home differ little except for the setting: in Afghanistan, the speech is a good-bye from the unit. At home, representatives of the armed forces and the political leadership usually pay tribute without the personal details common in ramp ceremony speeches (Hæren 2008a). What they emphasise instead are gratitude and expressions of indebtedness on behalf of the country. No distinction is made between soldiers and officers: 'This was not modern egalitarianism, but a long-established practice,' according to a British naval officer (Harnden 2011: 351).

The officer exaggerates the longevity of such traditions. Military units deployed and those staying at home recuperating or waiting their turn have always commemorated dead comrades, usually with a short speech by the officer in charge and a few minutes of collective silence. This practice continues, but repatriation ceremonies on the scale seen during the Afghanistan war are a recent innovation. Even in Britain, there were no procedures for receiving war dead as recently as the Falklands War (1982). Families were given the option of having their loved ones brought home or buried on the islands. A few opted for the latter. According to one widow, having the body brought back for re-interment was too unusual to be contemplated (Grice 2012). For those who wanted the remains of their family member repatriated, transport was effected quite unceremoniously, in shipping containers to a military port near Southampton. The remainder of the journey to the soldiers' final resting places took place in private hearses. There were no repatriation ceremonies or mourners respectfully lining the streets. Publicity announcing the arrival of the hearses was unheard of. Not until 2003 was a repatriation ceremony developed (Harnden 2011: 353). Its procedures bear a strong likeness to the US ceremony, from the pallbearers and the flag-draped casket to the folding of the Union Jack into crisp triangles before its presentation to the next of kin. Although no official explanation is available for the exact timing of the change, it seems the unpopularity of the Iraq War may have played a role. Saluting the noble qualities of the dead may have been seen as a way to counteract criticism of the war, though it is impossible to measure whether it had this effect.

The reception of coffins in Denmark has not been very different. Despite Denmark's less extensive military record, its practice actually antedates that of the British. Much of the ceremony performed for Danish victims returning from Afghanistan was introduced during the 1990s, after Danish peacekeepers were killed in the Balkans. However, close co-operation with the British armed forces since then has led to the adoption of at least one British tradition: a commemorative church service on Britain's annual Remembrance Day, in November, with participation of soldiers from both countries (Hæren 2008b).

For most of the war in Germany, the return of a dead soldier was mentioned only in the form of a communiqué announcing the transport to Germany. No details were given as to when the dead would be buried, let alone where. The possibility of the public expressing its sympathy, or indeed the opposite, was effectively precluded. From 2010, pressure from the media and the Bundeswehr rank and file led to greater openness and a clearer political acknowledgment of the loss of life. The turning point was a funeral for three soldiers in the same church, with remarks delivered by both the Chancellor and the Minister of Defence. Since then the German press corps has covered the funerals of dead soldiers in greater detail, and political attendance has been somewhat more noticeable as a possible result.

Still, as a rule, Danish and British newspapers have remained more generous about devoting space to such events.

Comparing the three countries, it becomes clear that public recognition of the dead is related in large part to the debate and information available throughout society. If the costs of war are addressed – including above all the emotional costs – the dead are received more openly. One should not overlook the positive effect of openness on next of kin and fellow veterans, who otherwise may have to endure emotional isolation. That was the clear message from an unusual number of surviving families that were personally profiled by Danish public television in 2008 and 2010 (Lyster 2008, Heilbuth 2010).

Although casualties in Afghanistan have been numerous compared with other armed conflicts since the end of the Cold War, the conventional wars of the twentieth century were far bloodier. The relative scarcity of deaths in Afghanistan should not be overlooked as a factor in public sensitivity. During World Wars I and II, grief and loss were universal. After the first days of the Battle of the Somme, a local British newspaper reported: 'There is hardly a home that has not experienced some great loss or suffered some poignant sorrow. … the grief and anguish of one of our number is the grief and anguish of us all' (Jalland 2010: 35). The psychological barrier between the families of the fallen and the rest of society would certainly have been lower at that time, partly due to the great number of deaths and partly to an attitude towards death and mourning that we no longer have. Pat Jalland, in one of the rare books dealing with World War II deaths, observes that in the decades following the end of the war, death more or less disappeared from public view (2010: 235). In the words of one doctor, death became not normal but horrible – it was no longer allowed: 'The bereaved are avoided with their grief bottled up … Somehow we manage to ignore death. We rarely see anyone die; most of us have never seen a corpse' (Lamerton 1973: 135). The families of the fallen often feel isolated with their sorrow, while soldiers who return alive feel that the barrier between them and the rest of society is unnecessarily high. The armed forces try in different ways to mitigate the worst effects. One form of help is often overlooked because of its voluntary, word-of-mouth nature: the practical assistance given by soldiers to the families of the fallen in their unit (Livgarden 2008). Performing such favours as house-painting, gardening and brick-laying demonstrates an attitude toward death and bereavement that differs from that of civilian society while keeping intact the unit's esprit de corps.

Naming the Dead

One way of diminishing the barrier that separates lost soldiers from society is to name them. If not, they only become part of the great mass of fallen. In Shakespeare's *Much Ado About Nothing*, Leonato asks the messenger 'How many gentlemen have you lost in this action?' who then replies 'But few of any sort, and none of name.' Afghanistan is different in that all the dead have been named. How important this has been for the other soldiers, is evident in all the books coming out of the War, naming plays a strong role.

The first official communiqué published shortly after death, contains few facts beyond the age of the soldier, his unit and where he was killed. Around the time of the communiqué's release, a personal tribute is generally given in the speech to fellow soldiers at camp, as the coffin is loaded for transport. The following quotation from a ramp ceremony is quite typical: '… he was always the first to make newer soldiers feel at home in the Company. … carry with you a picture of Jacko the man, the Geordie teenage who was a first-class soldier and a friend to all. Your duty is to apply his legacy with added resolve' (Pike 2008: 208).

Only after the next of kin is informed is the name published. At that point, the dead soldier is no longer just a young man killed by a roadside bomb, but an individual. Once coffins are received at home, the names of the fallen appear in official tributes. Strictly personal remarks have been unusual, with a notable exception for Søren Gade when he was Danish Minister of Defence from 2004 to 2010. His eulogies took the form of a personal message to the dead soldier, using the informal 'you' and praising his or her service to Denmark (Quist 2008). A photo of Gade teary-eyed during the ceremony assured him a prominent place in the newspapers. It also assured him a strong standing among the Danish rank and file in Afghanistan (Cordua 2009, Pedersen 2010).

Once a coffin is securely home, official information practices in Denmark and Britain give some leeway to speak of the soldier's personal characteristics. The web pages of the British MoD and the Danish Armed Forces have provided lengthy descriptions, not only of military careers but hobbies and interests (Hæren 2008a, 2009, MoD 2011). On the British MoD website friends and fellow soldiers, with the latter in clear majority, have added their own comments and reminiscences.

Again Germany remains an exception. The relative openness since 2010 has been restricted to the issues of who should participate on behalf of the government and where the funeral would take place. Politicians have probably pronounced the names of the dead in their eulogies at funeral services, but as a rule quality papers like *Frankfurter Allgemeine Zeitung*, *Süddeutsche Zeitung*, *Handelsblatt*, *Tageszeitung* and *die Welt* have not printed them. Those papers have rarely provided more information than what is available through official sources. *Bild*, a mass-circulation tabloid, has shown few such inhibitions. It generally covers the funerals in detail and is often able to present some background information on the fallen, filling in the meagre data available on the MoD home page. The difference between *Bild* and the rest of the media diminished when the funeral of the three Bundeswehr soldiers killed in April 2010 was covered. The Chancellor's presence in the church assured wider media attention to the event. Yet even though the names of the dead were mentioned in the tributes, most newspapers refrained from printing them, again *Bild* was an exception (2010). Soon after, the official German policy changed completely, when the Chancellor presented the government's Afghanistan policy in the Bundestag (Merkel 2010b). Her speech opened with the names of the three dead, followed by the names of other Bundeswehr soldiers who had lost their lives that year. As far as can be ascertained, nobody criticised the Chancellor for the change; on the contrary, in the media and on a soldier blog alike it was praised as an important step towards acknowledging the dangers confronting the men and women on the ground (Hautkapp 2010). Yet, she did not attempt to individualise the dead in any way by referring to their career, background or personal characteristics.

In fact, the only time the names of war dead were used as part of the political debate in the Bundestag occurred when MPs from the Left held up sheets of paper with the names of civilians killed in the Kunduz bombing (*Spiegel* 2010). The spokesperson of the faction claimed the action was motivated by the Bundestag's failure to commemorate the civilian victims with a special ceremony. In Britain, too, the dead were named by war opponents. In one of the country's few demonstrations against the Afghanistan war, the names of fallen soldiers and civilian Afghans killed were read out by activists standing beneath the London Cenotaph's inscription '*Our Glorious Dead*'. According to one of the campaigners, in order to bring home the costs of war to the public it was necessary to individualise the casualties by using their names (SKY 2009).

Germany: No War = No Fallen

When the first dead soldiers were brought back to Germany in 2002, the official declarations were brief. They nevertheless had greater impact than the official communiqués of the British and Danish authorities because, perfunctory or not, they constituted the final word said. As mentioned, in Britain and Denmark the initial communiqués were quickly followed by a description of the fallen published on the web. In Germany, this was never done. Until 2010, Germany's dead were not publicly named and nothing was known about them. Without a face and a name, let alone a brief CV, the public could not form an impression of the soldier. The impact this had on the debate about the war was easy to observe. Instead of dwelling on casualties, political focus was directed to the perceived break in the country's modern history as a civilian power.

A former Bundeswehr officer, Andreas Timmermann-Levanas, has pointed out that the death toll conflicted not only with the framing of the war as an assistance project, but also with the image of the Bundeswehr as an instrument of peace. There was no room, he writes, for the dead (Timmermann-Levanas and Richter 2010). More strictly speaking, there was room for death, but not for the fallen. This was evident from how causes of death were classified on the German MoD's homepage (Müller 2012). While British and Danish official surveys have been sufficiently detailed to give a clear impression of whether the cause of death was combat or accident, such information was withheld in Germany. The German website simply listed *Durch Fremdeinwirkung* [due to external causes]. This is a wide category ranging from combat to roadside bombs. The only specific cause of death to be categorised separately was suicide. Indeed, while only 19 German soldiers have taken their own lives while deployed outside Germany since 1998, the website leaves the impression that suicide is a major cause of death. If other causes had been specified – roadside bombs, for instance – debate might have arisen over such issues as the thickness of vehicle armour. But political leaders and the public had no such information. As late as 2007, when 20 soldiers had lost their lives, it was discovered that not even the Defence Committee in the Bundestag had been informed about the exact causes of death. It relied instead on the all-inclusive categories mentioned above (Jörges 2007). As long as it was close to impossible for outsiders to decipher how German soldiers were dying, the government had an easier time denying it was involved in a war.

In this, the government was abetted by German commanding officers in Afghanistan; as far as can be ascertained no prominent German officer has come forward while still in service to contradict the official framing of the war. A good example of how the official line has been echoed on the ground is an interview given by the newly appointed ISAF Regional Commander Markus Kneip to a leading German newspaper as late as 2011 (Meyer 2011b). He stated that while every loss was painful, his feelings would be exactly the same whether a soldier was hit by a Taliban sniper or by a car when crossing the street at home. His comparison made all casualties equivalent, and thus obscured conditions on the ground. It was an attempt to reduce the political potency of casualties: if all the dead were mourned equally, there would be no reason to focus on soldiers killed by the Taliban.

Such an attitude, if made official, would have consequences for the families of dead soldiers. Widows, for example, receive a war pension only if their spouse is killed in war (*Stern* 2009). A minor but strongly symbolic consideration is that the maintenance cost of a war grave is covered by the German state (Junge 2009). Soldiers wounded in war, meanwhile, can claim special state financial and medical support. Not until August 2011 was this granted to Afghanistan veterans, after considerable pressure from the German Federal Armed Forces Association [BundeswehrVerband] (Bundeswehr 2011). Until then, when Afghanistan was on the Bundestag agenda, Bundeswehr casualties had never been debated in depth. Related issues like the numbers of wounded and traumatised returnees were also avoided. As time passed the silence caused considerable criticism outside parliament (Tenberg 2009). Especially the Armed Forces Association and the Association for Reserve Officers [Reservistenverband], grew impatient with the politicians' reluctance to address the human costs of war. The media shared their impatience, repeatedly focusing attention on the lack of political debate and producing a growing number of documentaries and TV dramas.[3] In fact, one of the few times the Minister of Defence referred in parliament to

3 Among the most significant documentaries are the series *Streitkräfte und Strategien* (NDR) which has focused repeatedly on Afghanistan since 2005; and the six-part documentary *Operation Afghanistan – Die Bundeswehr im Einsatz*, screened in late 2007. Other notable films piercing the official silence were the TV dramas; *Nacht vor Augen*, 2008; *Willkommen Zuhause*, SWR 2008; *Tatort: Heimatfront*, ARD, 2011.

the psychological impact of war, his source of information was a television play dealing with the mental problems of an Afghanistan veteran (Meyer 2009: 29).[4] Although the lack of hard facts on the extent of the problem should have surprised the Bundestag, the response by veterans was in any case positive (Dirk 2009).

The reluctance to debate the war was consistent with the Chancellor's absence, until 2010, at repatriation ceremonies and funerals alike (*Handelsblatt* 2010). Her failure in particular to appear at funerals became a source of criticism, with the *Frankfurter Rundschau* publishing a scathing critique in September 2009 posing the question if she cared at all (*Frankfurter Rundschau* 2009). Although high-ranking representatives were always present when the planes carrying coffins landed, and senior officers were present in church when religious ceremonies took place, the Chancellor's own attentiveness was widely perceived as insufficient. Her non-participation was condemned by Social Democrats who supported the government's Afghanistan policy as well as by the press and Bundeswehr organisations (*Merkur* 2010). When the three Bundeswehr soldiers were shot by a Taliban member, in April 2010, a long while passed before it became clear that the Chancellor would participate in the funeral. Even then, the Chancellery initially emphasised that she would go as 'a private matter' (Gebauer 2010). The only time she had attended a similar ceremony had been during a visit to Kabul, and in that case the dead soldier had once been one of her bodyguards.

The biggest change in Germany around this time was not the way funerals were arranged or even the possibility that the Minister and the Chancellor would attend. Those changes did affect how German soldiers killed in Afghanistan were to be regarded. Underlying, however, was the far more complicated issue of how Germany's deployment should be framed. As long as it was framed as an assistance mission, the soldiers killed in it would represent tragic accidents; they could be portrayed as honest and worthy representatives of Germany, but little more. Official declarations would emphasise their altruism. As providers of help, they embodied the most lauded values of post-war German political culture. The fact that they were soldiers on a military mission was not mentioned until July 2009, when Defence Minister Jung gave the eulogy for three Germans killed by the Taliban and described them as 'good soldiers and true patriots' (Jung 2009). Still, the official framing of the mission made it impossible for him to take the final step and describe the dead as fallen. This was not only a matter of 'tactical semantics', as a senior MoD civil servant claimed (FAZ 2008). It had profound psychological consequences for those serving in Afghanistan, who soon discovered that what they endured was officially denied.

In April 2010 this changed when the Chancellor and the Minister of Defence read eulogies in church for the three men killed in Afghanistan. The Chancellor ended with the words: 'I bow my head for you. Germany bows for you' (FAZ 2010). This was an unusually emotional statement from a politician known for

4 The film in question was *Wilkommen Zuhause* [Welcome Home], produced in 2008.

her restrained public demeanour. For that reason, the words were on virtually every front page the following day. Politically, her speech was important because it broke with the strong German inhibition, not to say taboo, against linking national gratitude with dead soldiers. In previous statements made in the Bundestag, casualties had usually been communicated with 'profound sorrow' and little more. The importance of the new tone was momentarily overshadowed by the explicitly political contents of the Chancellor's speech. She said Germany was not going to withdraw until the mission was accomplished. Moreover, she expressed understanding for the soldiers' use of the term 'war' to describe what they were experiencing, adding, 'and I can well understand them' (FAZ 2010). Defence Minister Guttenberg went a small step further, stating that the soldiers 'for understandable reasons use the term war – so do I'. Guttenberg expanded on the Chancellor's theme of national gratitude. The dead, he said, are 'our country's brave heroes', inspiring 'pride' (Soldatenglück.de 2010). Such praise had not been heard before. Equally surprising was the almost uniform praise that greeted the Chancellor's speech. One might have expected the press to attack her for nationalism or reinvented militarism, but apart from some comments in communist newspapers such as *Neues Deutschland*, the response seems to have been characterised by a sense of relief that the government would finally call it a war. Only some protests were forthcoming from the German Peace Union [Deutsche Friedensgesellschaft], which had been running a campaign in favour of a unilateral German withdrawal from Afghanistan (*Neues Deutschland* 2010).

Not long after participating in the funeral of the three men, the Chancellor outlined the government's Afghanistan policy in the Bundestag (Merkel 2010b). Her justification for the deployment and summary of what Germany wanted to achieve contained little that had not already been said. The only section of her speech that was a radical departure concerned the dead. This time, she gave a detailed account of who had accompanied the bodies back from Afghanistan. She explained what the Minister of Defence would do to express his support for the troops stationed there, and designated which government ministers would attend funerals. Echoing her church remarks, the Chancellor also said: 'When most of the soldiers describe what they experience in Afghanistan on a daily basis as civil war, or simply as war, I can well understand them' (Merkel 2010a: 3). The fact that the government had so far refrained from using the word, should not, the Chancellor underlined twice, be interpreted as an attempt to play down what the soldiers were going through. When her declaration was finished and debate ensued, the Bundestag missed the opportunity to discuss what the deaths in Afghanistan meant to Germany. The triple funeral was still on everybody's mind, but apart from one member's cautionary remark that it was important not to get used to casualties, not a word about the deaths was uttered (Merkel 2010c).

The Bundestag's passivity did not change. What did change was the use of funerals as an arena to defend the government's Afghanistan policy. With media coverage assured whenever the Chancellor was present, the newspapers and the television began paying more attention to what was said about Afghanistan at funerals than in parliament. The funereal setting, with its flowers, coffins and mourning audience, provided a more emotionally charged backdrop than the plenary hall of the Bundestag. The message itself was enhanced by images of the always-packed church and the additional numbers gathered outside. When three men were buried in the small Bavarian town of Regen in February 2011, about a thousand people gathered to follow the ceremony on a big screen outside the church (*Süddeutsche Zeitung* 2011). The Chancellor participated, albeit without any of the media buzz her attendance had created the preceding year. She did not speak, but her newly appointed Minister of Defence, Thomas de Maizère, paid his respects to the three dead soldiers in unexpectedly political terms (*Westdeutsche Zeitung* 2011). He defended the policy of the government and reiterated his confidence in its strategy. The important work done by the soldiers should be given more support by the population, the Minister concluded. His admonitions were aimed not only at the general public, but at Claudia Roth, leader of The Green faction in the Bundestag. She had wanted to use the deaths of the three soldiers and a number of other attacks that occurred around the same time as grist for a debate over German unilateral withdrawal (Gebauer et al. 2011). The Minister chose to reply at the funeral rather than in parliament or in the press. That way he was assured full attention.

One element that remained unaffected by the political changes was the hymn sung at most funerals: *der gute Kamerad* [the good comrade], a sad song about the loss of a 'good comrade' dating from the Napoleonic wars. It is popular not only among German soldiers but also with the French Foreign Legion. Danish soldiers are familiar with it, too. One Dane writing about death in his unit in Afghanistan used the line *Ich hatte ein Kamerad* [I had a comrade] as a chapter heading.[5] The song's theme is the reason for its popularity. The arbitrariness of death described in so many of the soldiers' accounts is expressed in the second verse:

> A bullet it flew towards us,
> For him or meant for me?
> His life from mine it tore,
> At my feet a piece of gore,
> As if a part of me.[6]

It is sung or played at both official occasions and private funerals. In Afghanistan, too, the tune is played when a dead German soldier is sent back home (Groos 2011b: 112–3). Those who have come back from Afghanistan and recounted their experiences in battle have, almost without exception, expressed the same

5 This is a misquotation from the song, whose actual line is "Ich hatt' einen Kameraden', see Birkeland 2010: 276.

6 Translator Frank Peterson (n.d.), The German original goes: Eine Kugel kam geflogen/Gilt's mir oder gilt es dir?/Sie hat ihn weggerissen,/Er liegt zu meinen Füßen/Als wär's ein Stück von mir.

feeling of shock and sorrow reflected in the song (Schindler 2010). A contrasting interpretation has been proffered. The historian Alexandra Kaiser sees the song as a portrayal of how military hierarchy and obedience prevents empathy in that the protagonist is so busy loading his rifle, that is obeying orders that he refuses to fulfil the dying man's last request (Kaiser 2010: 84).[7] Had other shared her interpretation, the song would not have been such a rare example of historical continuity in a country whose recent military priority has been to frame the Bundeswehr not only as a break from its militarist past but in many ways the antidote to it. Because its lyrics fit the official pacifism of the post-war ear, the song has been used in all official commemorations of the fallen in the World Wars.

Britain: Fallen Heroes

In these pages, the description of the German attitude towards the war dead has focused on the role played by official representatives in funerals. Popular participation has not been addressed for the simple reason that it has remained minimal, restricted to local inhabitants attending the church service or gathering outside the church to pay their respects. Britain is different. There, popular participation has been an important part of the homecoming and is actively encouraged. The best known and undoubtedly the most photographed example of popular participation remains the line-up of inhabitants in the small town of Royal Wootton Bassett as they paid tribute to hearses passing through their streets. It has been quite a frequent sight because the town is situated on the route between the air base where the planes land with the coffins and the hospital in Oxford where the coffins are transferred pending funeral. Before they reach the town, a formal repatriation ceremony is held at the base: 'The military service, with military pallbearers and volleys of gunfire over the coffin, is very moving' (Danziger 2010: 6). Attendance there is restricted to next of kin.

However, sometimes family members have preferred to stand together with locals on the pavement instead of participating in the ceremony at the RAF base (Allen and Savill 2008). The local branch of the British Legion, the ex-servicemen's association, has made it known that political speeches are not welcome. As a result the ceremony has been kept free from attempts to use it for official or political purposes. This was put to the test in 2009 when a radical Islamist group called Islam4UK declared it was going to stage a demonstration to attract public attention to the plight of Muslims in Afghanistan. The ultra-nationalist English Defence League then announced that it planned a counter-demonstration. After considerable pressure from the police and local authorities, both sides cancelled their plans. Since then the ceremonies have remained remarkably free of political attempts to exploit the town's way of showing respect.

---

7 The verse in question is the following: His hand reached up to hold mine. I must re-load my gun. [Will mir die Hand noch reichen, Derweil ich eben lad] (Peterson n.d.).

In 2011, the town was granted the right to use the prefix Royal by the Queen as an official sign of gratitude (College of Arms 2011). The new status was not welcomed by all the townspeople. The mayor stated that he thought the inhabitants did not want any particular recognition, that their way of honouring the dead should not seen as anything more extraordinary than an expression of gratitude (Wallin 2009). In any case, the nearby air base was closed down in September 2011, ending the funeral procession though Royal Wootton Bassett. Since then the planes bearing the dead have landed at another air base. The route of the hearses has been changed, too, no longer passing through Royal Wootton Bassett but through the village of Carterton. And there the tradition of paying respect has continued. To facilitate people's attendance, repatriations have been announced on the local county website (Oxfordshire County Council 2012). In addition, The Royal Air Force Police Association (RAFPA) has maintained a repatriation website (RAF 2011). The names of dead soldiers, their military rank and affiliation have been provided along with information pertaining to the circumstances of their deaths. The information has been based on details made available by the MoD. However, in contrast to the ministerial page with its factual tone, the RAFPA has put the casualties into a historical context. When the page is opened, a wreath of red poppies meets the eye. Originally worn in remembrance of the World War I dead, the poppy has since become the symbol of all of Britain's fallen soldiers. Red poppies appear almost on every website, private or public, that commemorates a dead British soldier. Underneath RAFPA's wreath is the couplet

> We will remember them
> We will remember them

taken from 'For the Fallen' by Laurence Binyon.[8] The lines are often repeated during memorial services, and are often printed on ribbons attached to the poppy wreaths placed on war graves.

Royal Wootton Bassett and Carterton have become well-known because of the frequency of repatriation events. Both in Germany and Denmark, members of local communities also pay their respects when one of their own is brought home to rest. The two British towns stand out because the passing coffins contain men whom the locals do not know; their communities just happen to have been on the quickest route from the base. The emotional distance that so many veterans complain about on their return home, after noticing that many people fail to grasp what the war is about, is less pronounced in Royal Wootton Bassett and Carterton, where the armed forces play a highly visible and integrated role. But it

8 On the RAF Police Association website, Rudyard Kipling's *Tommy* is mistakenly given as the source. This was a poem written by Kipling in the 1880s about the lack of appreciation shown to soldiers by the public. Kipling wrote *My boy Jack*, a moving poem, and hugely popular at the time, about the loss of his own son, who went missing in September 1915 at the Battle of Loos.

would be too easy to reduce the behaviour of residents there to this factor alone. The passing of the hearses was also an opportunity for ordinary people without military ties to express support for the troops. That such a need is prevalent outside garrison towns was shown by the 'Heroes Welcome in the UK' initiative, which began in Scarborough, a town with no military links. Initially, posters and stickers were printed up so that shops, theatres and amusements parks could signal that servicemen and women would be particularly welcome, and in some cases given a discount. It was launched as a response to the reports of problems faced by returning soldiers, especially their sense of not being appreciated (Davies et al. 2008: 39–40). Soon the campaign spread all over the country, and local councils joined as well.

That campaign and the public gatherings to honour returning coffins should not automatically be considered synonymous with support for the war in Afghanistan. On the RAF Police Association blog the focus, in descending order, is the dead soldier, the armed forces, and finally Afghanistan. The refusal of Royal Wootton Bassett inhabitants to condone any form of political statement was an acknowledgement that the dead represent different things to different people. In Parliament, fallen soldiers have been a part of the political debate, but when their bodies are received, transported and buried, their role changes. First of all, they become less politicised. Even though political leaders have regularly attended the funerals, and have given the eulogy on some occasions, they rarely express anything beyond admiration and gratitude. The importance of the Afghanistan mission to the protection of Britain and the improvement of life for Afghans is sometimes mentioned as well, but the political message is usually general and rarely controversial.

Dead Danes

The previous chapter outlined how the Danish government changed its framing of the war and chose to focus increasingly on the men and women deployed rather than the mission's goals. This change was reflected in how the dead were received and buried. Although the original aims of regime change, stability and development were not completely forgotten, the soldiers and their sacrifices began to loom larger. The altered rhetoric is evident when comparing speeches made at the beginning of the war against later tributes paid to the dead by the Prime Minister on Denmark's Memorial Day (5 September) (Rasmussen 2009, 2010, 2011a, 2011b). Over time, the personal qualities of the dead have come more and more to the fore. One reason for the change in focus has been the diminishing prospect for victory in Afghanistan. When the first casualties occurred, it was possible to link them directly to the fight for security both in Denmark and in Afghanistan; as the end of the war nears, continuing to insist on such a link would suggest an odd disregard for developments on the ground.

Focusing on the soldiers' military and personal virtues has fitted another of the government's projects: overcoming the pacifism that characterised Denmark's security policy in the final decade of the Cold War. One way of doing that was simply to create a historical bridge between the current deployment and past Danish military exploits and heroes. Positively loaded words became key elements in this bridge. Government leaders, journalists and security experts began referring to the Danish soldiers in Afghanistan as freedom fighters, a term that traditionally had been reserved for the Danish resistance movement of World War II (Rasmussen 2008, Ruus 2009, Rottbøl 2011).

Naturally, the new political framing corresponded with that of the soldiers themselves. This neat alignment of perspectives was evident in the choice of a hymn for base repatriation ceremonies and church services: *It was a summer morning* [Det var en sommermorgen] composed after the Danish army defeated the advancing Prussians at the Battle of Isted in 1850. Despite the pacific-sounding title, the second verse (the one that's always sung) is strongly patriotic:

> for the honour of old Denmark
> the Danish people fought
> yes for the Danish tongue
> and for the Danish land.[9]

The Danish hymn is a jubilant celebration of national values and the need to defend them, arms in hand.

By coincidence in time, fundamentalist Muslims in Denmark unwittingly came to assist the government in its shift to glorifying Danish soldierdom. In 2006 the so-called cartoon crisis erupted after a Danish newspaper printed caricatures of the prophet Muhammad. Relations between Denmark and Muslim countries deteriorated rapidly. Danish embassies were looted and effigies of the Danish Prime Minister were burnt together with the Danish flag. For months, Danish television news and newspapers carried stories indicating a steady worsening of relations throughout 2007 and into the following year. In 2008, a Taliban commander declared that the cartoons were reason enough to attack, kill and ultimately expel Danish troops from Afghan soil (Brix 2008).

The extent to which the cartoon affair influenced popular perceptions of the Danish deployment is not easily measured. The soldiers themselves rarely mention it in their accounts of service in Afghanistan (Hundevadt 2008: 246). However, the government's framing of the crisis emphasised the need to stand up for Danish democratic ideals and values and against a totalitarian threat (Hansen 2011, Jensen 2012). Repeated attempts by Muslim terrorists to kill the cartoonist and blow up the newspaper building where the cartoons had been published served as a reminder of the threat for years to come.[10]

---

9 In Danish, the lines quoted are: For gamle Danmarks ære/da stred det danske folk/ ja, for den danske tunge/og for det danske land.

10 In 2008, an attempt was made to kill the cartoonist. The following year, a US–Pakistani man was arrested in Chicago after plans to blow up the newspaper that had printed

In early 2011, the official framing of fallen Danish soldiers as defenders of democracy and freedom was further abetted by the Danish branch of the islamist Hizb ut-Tahrir (Party of Liberation). This small grouping, with a core of approximately 50 members, had attracted attention over the years for its extremism (Grøndahl et al. 2003: 116). One of its aims is to call all Muslims to holy war, or jihad, against Israel and the West. It regards suicide bombers as martyrs. The Western forces in Afghanistan are classified as occupiers and thus legitimate targets. Several European countries, including Britain and Germany, have imposed restrictions on the group's activities. In January 2011 the organisation arranged a public meeting in Copenhagen's Royal Library, a prominent building near the parliament. Posters announcing the event carried photos of coffins draped in the Nordic flags and carried by soldiers. The caption ran: 'Afghanistan. The Scandinavian governments in the service of the USA'. There were also statements lauding the armed resistance against Western invaders and praising the killing of Western soldiers (*Jyllands-Posten* 2011). The Danish veterans' organisation The Blue Berets [De Blaa Baretter] as well as The Central Association of Permanent Defence Personnel [Centralforeningen for stampersonel] both reacted with outrage, claiming the statements were tantamount to treason (BT 2011, *Politiken* 2011b). Attempts were made to have the meeting stopped. These failed, with the Minister of Justice arguing that freedom of expression applied even to Hizb ut-Tahrir (*Jyllands-Posten* 2011). A counter-demonstration was then arranged outside the Library featuring political leaders, Afghanistan veterans and civilians.

The meeting was hardly a turning point in the framing of the war. If anything, it galvanised public support for the soldiers. Members of the governing parties used it to clarify for Danes at home the kind of enemy their forces were fighting in Afghanistan – the kind of men, in other words, who caused the death of Danish soldiers (Lindquist 2011). This had been the official line in all the Prime Minister's new year's addresses held since the war started. During the cartoon crisis, the demonstrations in Muslim countries, sacking of embassies and flag burning were all described as the result of totalitarian forces threatening Denmark and Danish values (Rasmussen 2007). After encouraging all to guard these values, he moved directly to thanking the Danish soldiers deployed abroad for their service to Denmark.

'He died a hero' – Exceptional Deaths

At the official level, the response to the death of a soldier in Afghanistan has remained basically unaltered throughout the war in Denmark and Britain. In

the cartoons were discovered. Similar plans led to two separate arrests in 2010. That same year, the cartoonist was attacked in his home by a Muslim immigrant. Also in 2012, arrests were made in Norway and in Northern Africa of men planning to kill the cartoonist.

Germany, the only changes have been higher-profile political participation, increased media coverage and a new terminology for describing the war and the dead. Otherwise, dead soldiers in all three countries have been repatriated and received in the much same way since the first death in 2002. However, some casualties have been granted more attention than the rest – not because of any official decision to idolize certain individuals, but because soldiers at times have reacted spontaneously to honour a fallen comrade they liked and admired. Once such an effort has attracted media attention, the dead assume a political status nobody could have anticipated.

In Denmark, Anders Johan Stæhr Storrud became well known after his death in October 2007 because fellow soldiers stepped forward to tell of his courage and leadership. In Britain, the death of British Lieutenant-Colonel Rupert Thorneloe in July 2009 caught public attention for similar reasons, in addition to the fact that he was the first senior officer to be killed since the Falklands War (*Telegraph* 2009a, Tran 2009). Storrud died after being hit in the head by a bullet. His death was not instantaneous, and the first news bulletins merely stated that a Danish soldier had been gravely wounded. Only after his family was informed of the death was the name released to the press. The initial media coverage contained nothing exceptional. Testimonials from soldiers on the Internet provided the first signs that Storrud had been an officer out of the ordinary (Mindet 2007). The posted tributes were heavy with a shared sense of loss (Hundevadt 2008: 172–6, Caruso and Stougaard 2009, Garderportal 2009, Birkeland 2010: 207–8). The dead officer's motto – *Today we mourn our fallen, tomorrow we'll attack again* – was quoted as proof of his attributes. The religious service was covered in great detail on the armed forces website, the priest's eulogy was transcribed and the attendance of the British and US defence attachés was noted (Hæren 2007). By the time of the funeral, media attention was pervasive (TV2 2007). One reason was Storrud's own personal development. Having grown up with a pacifist mother and a communist father before deciding on a military career, he embodied Denmark's security policy transformation in the post-Cold War years.

The only one to question Storrud's sudden fame was his widow (Høy-Jensen 2008). She warned against idealising his death, because doing so would gloss over the problems many officers had adapting to civilian life on returning home. Her warning was barely noticed, and Storrud's name and merits remained points of reference and admiration not only for his fellow-soldiers but for the government. The Minister of Defence hailed Storrud in heroic terms after a newspaper cited his death as an argument for swift withdrawal. In the article 'Mission Impossible. Another life wasted', the paper portrayed Storrud's death as the consequence of a foolhardy attempt to recover a stranded vehicle (*Ekstrabladet* 2007). Danish soldiers in Afghanistan reacted to the article with bitterness; a conservative MP wanted the paper boycotted (Selchau-Mark 2007, Hundevadt 2008: 176). The Minister of Defence gave an interview in which he called Storrud's attempt to recover the vehicle an act of sacrifice (BT 2007). In the hands of the enemy, the vehicle could have become a formidable weapon against Storrud's own men. The recovery attempt had been the result of Storrud's own decision, taken on the spot to save the lives of the others, the Minister stressed (Worm 2007). This effectively silenced the debate while supporting the government's framing of the men and women in Afghanistan as outstanding individuals.

The circumstances surrounding the death of Rupert Thorneloe were similar to Storrud's, but the repercussions were different. The day he was killed, eight other British soldiers suffered the same fate, making it the bloodiest single day for British combat troops since the Falklands War. Despite the general carnage, Thorneloe's death received far more attention than the other casualties. The attendance of the Prince of Wales and the Prime Minister at Thorneloe's funeral increased the media focus even further (Morris 2009). The Prime Minister praised Thorneloe for 'his unquestionable loyalty', while others underlined that he had gone to Afghanistan 'to win the hearts and minds of the Afghan people' (*Telegraph* 2009c). Both statements fitted the official British framing of the war. A letter that Thorneloe wrote to the mother of a soldier killed weeks before his own death found its way into print, showing him to be not just an inspiring officer but a compassionate leader (Channel 4 2009). His exemplary qualities made him a convenient model and reference point for both the government and the opposition. Probably no other member of the British forces killed in Afghanistan has been mentioned as often in Parliament.[11]

Not all such references were in tribute. Many focused on the cause of Thorneloe's death, a roadside bomb, and used it in political debate. Shortly before his demise, Thorneloe had written a secret memo warning that Britain's shortage of helicopters meant more transport by road, and thus unnecessary exposure to IEDs. The warning was repeated just weeks after Thorneloe's death in a report by the House of Commons Defence Committee (2009: 50). The loss of Thorneloe reinforced the impression that insufficient funding was costing lives. In the media, parents of men who had been killed by roadside bombs criticised the Prime Minister, and at one point Thorneloe's father joined the critics (*Telegraph* 2009b). In Parliament, one MP argued that Thorneloe's death could have been avoided had the government not mixed 'strategic incompetence with political cowardice' (Allen and Groves 2009, Davis 2009). Late in 2009, the BBC reported that new and more robust helicopters had arrived in Afghanistan; the pressure generated by Thorneloe's leaked memo was cited as a main reason (BBC 2009).

In Germany, no casualty has generated so much attention, debate or admiration. It's not that Germany has lacked men of similar virtue; the country has just been reluctant to underline the traditional skills associated with warfare. General Kneip, mentioned above for his refusal to differentiate between causes of death, was asked if he would single out any of his soldiers as particularly outstanding. His reply was that 'all my soldiers are everyday heroes' (Meyer 2011a). That conformed

11 According to Hansard, in the House of Commons his name was mentioned four times on 8 July 2009, four times on 16 July and twice on 2 November. In the House of Lords his name and those of other fallen soldiers were read out on 9 July 2009, and he was mentioned again on 14 October and 6 November the same year.

to the framing of the war itself and of the Bundeswehr soldiers as 'civilians in uniform'. Yet, by this time the general's egalitarianism had already been weakened by the institution of bravery medals. The first were awarded in July 2009 by the Chancellor. The decision to issue them earned her vehement protests from the Left and from peace groups outside the parliament. To them, the medals were a clear sign that the government was conducting a slow remilitarisation of German security policy (dpa 2009, Bergmann 2011). Loud as they were, the opponents remained a clear minority. With the exception of the Left, no political parties criticised the awarding of bravery medals. On the other hand, no party came out in support of them either, though a few conservative parliamentarians did speak up. The main reason for the silence in parliament was uncertainty over how the public would react. The same uncertainty accounted for strong restrictions imposed on media coverage of the award ceremony. Press coverage of award ceremonies in Denmark and Britain has been strong (MoD 2009). In Germany only a few minutes were set aside for journalists, whose invitation was conditioned on an agreement to confine their questions to one soldier, hand-picked by the Bundeswehr, who would meet the journalists in a separate room (Schuler 2009). The ceremony itself was conducted in the privacy of the Chancellery. With no public ceremony, speeches or military music, the government hoped to avoid being seen as idealising the armed forces and the soldier in question.

Less than a year later, when the three soldiers were buried at once, official restraint weakened. In Minister of Defence zu Guttenberg's tribute to the three, he asked rhetorically whether they were heroes and whether one should feel proud of what they had achieved. Both questions, he said, could be answered 'not politically, but simply with yes' (*Bild* 2010). Because the soldiers had been killed by the Taliban, they were praised in their capacity as soldiers on a combat mission. Reactions in the media and within the military suggested approval, if not relief. While Merkel had lauded the recipients of the bravery award in the privacy of her office, zu Guttenberg used a public funeral to deliver the same message.

Conclusion

The dead have come to play widely different roles in the public discourse of the three countries examined here. The least change over time was observed in Britain, whose dead from Afghanistan automatically join the long honour roll of fallen British soldiers. Although the solemn receptions at air bases are a fairly recent addition to British rites, it is hard to find anybody who opposes them. The war may be criticised, but not the dead. The calling out of names of lost British soldiers and Afghan civilians during an anti-war demonstration in London was intended to link both groups as victims, not to belittle the soldiers' status. Furthermore, the gesture was meant to protest the Afghanistan war, not the principle of deploying British soldiers outside the UK. When, by contrast, Germany's Left faction held up the names of Afghan war victims on posters in the Bundestag, it was signalling its opposition to sending the Bundeswehr abroad. To the Left, there was little apparent difference between the Afghan war and the aggressive endeavours of the Third Reich. In both cases, civilians have had to pay the highest price.

The Left's display of Afghan names was an attempt to counter the image of German soldiers as 'civilians in uniform'. Less explicitly, it portrayed the mission in Afghanistan as war – the opposite of what the government has claimed through its framing devices. Only when the official view became untenable, as casualties rose, did the government's rhetoric change. That the change occurred at a funeral, to be followed up with a policy declaration in the Bundestag a few days later, was politically opportune. The Chancellor's policy address to parliament contained the same message, but its political contents had already been defused by her speech in the church.

While Merkel's speech was unexpected, the receptions and burial ceremonies of fallen Danish soldiers were integrated into a longer-term process of political change. Denmark's dead from Afghanistan have become a bridge to a distant past when Danes fought and died for national values. That Denmark's nineteenth-century wars against Prussia were primarily about national survival has not deterred Danish politicians or soldier-authors from declaring them relevant to the Afghan mission. Objections to such narrative framing have been minimal. Nor has opposition to Danish participation in the war amounted to much, suggesting widespread acceptance by the public. It goes without saying that the linkage to past military glory is very popular with soldiers. Their image as courageous representatives of Denmark gains potency whenever someone is killed. Each death is seen as a sacrifice on behalf of Denmark, a point made abundantly clear in official speeches and other formalities surrounding the coffin reception and in the choice of hymn for the funeral. Such an approach ensures that the costs of war are never glossed over. Depicting men and the woman killed by the Taliban as family members, not just soldiers, gives a more credible account of 'civilians in uniform' than Germany's practice, in which the authorities provide little background information and the press rarely interviews the next of kin.

The Danish way of repatriating, receiving and burying the fallen has differed little from that of the British; the resemblance is so strong that it seems likely the Danes modelled their practice on British procedure, much as the British modelled theirs on the American. As in Denmark, the absence of critical voices in the UK may be taken as an indication that most people, including soldiers and their families, regard the established procedures as appropriate official recognition. The spontaneous displays of respect from the inhabitants of Royal Wootton Bassett and Carterton represent a still more powerful form of expression. One British officer commenting on the displays suggested that they might stem from a 'Princess Diana effect' (King 2010: 18, Wilson 2009–2010). Although that explanation cannot be entirely discarded, the example of Antigone in Sophocles' tragedy shows that honouring the war dead is an ancient phenomenon.

Before ending, a few comments should be made on how the reception and burial of dead soldiers influence casualty sensitivity and ultimately support for the war. It cannot be said that the presence of high-ranking officials at public ceremonies either diminishes sensitivity or boosts support for the war. We could safely draw that conclusion if support for the war in Britain had remained as persistently strong as it has in Denmark, or if Germany's attitude to the war had turned any less negative after the Chancellor's appearance at the April 2010 triple funeral. The way of organising receptions and funerals has had some impact, but much depends on the framing and the extent to which it seems credible. The German policy of withholding cause of death, restricting information on upcoming repatriation ceremonies and keeping top political leaders out of church eventually seemed so disconnected from events in Afghanistan that the government had to adapt. In Britain, criticism emerged over what was seen as a gap between political declarations praising the armed forces and a perceived failure to honour the dead. Yet at no point have the accusations levied against the government questioned the legitimacy of the war itself.

In Germany, political support for the deployment was weak from the outset. This effectively ruled out any debate in the Bundestag about Bundeswehr casualties. Nevertheless, as we have seen here, the speeches by top political leaders in church marked a transition towards greater openness. It is unlikely that the more open approach has affected public support for the war; there is no opinion poll to that effect, in any case. One reason may be that the framing of the war as one of help and assistance remains too entrenched in popular opinion. The Chancellor and the Defence Minister certainly adjusted the frame at that 2010 triple funeral in Selsingen, but the original rationale for deployment has never been abandoned. Moreover, the revised official rhetoric, although perceived as closer to the facts, may have done little to boost the government's credibility because it contradicted so much of what had been said before.

The Danish and British governments entered the war both to help the civilian population of Afghanistan and to remove the Taliban by force. High numbers of casualties may not have been expected, but they were never ruled out, as in Germany. Moreover, using the armed forces to maintain international influence (in Britain's case) or to gain it (in Denmark's) had been an openly declared policy long before the Afghanistan war. In Germany, the ability to deploy the Bundeswehr outside the country to confront security threats had been part of official policy since the 1994 Bundeswehr *Whitebook*, but that was more a case of tacit agreement than openly debated policy (1994, Martinsen 2010). The lack of an explicit public acceptance of war deaths caused new problems when public discussion about a war memorial started towards the end of the war. About the same time, a decision was made in Denmark to create a war memorial. In all the countries analysed, a three-way balance had to be found between consideration for the mourners, public recognition of the dead and the overall framing of the war. This will be discussed in the next chapter.

Chapter 4
Commemorating the Dead

When a British soldier was repatriated from Afghanistan, the following lines from *For the Fallen* by Laurence Binyon were read out: 'They shall not grow old as we who are left grow old. Age shall not weary them, nor the year condemn. At the going down of the sun and in the morning we shall remember them.' This chapter is about that pledge. Official efforts to honour it have resulted in public commemorations of war dead in the form of national memorial days and monuments. These are official acknowledgements of loss that bridge between the private grief of a dead soldier's relatives and the gratitude of society. Most families of the fallen want their loved one's death to be publicly recognised and remembered. In all three countries examined here, sacrifice in war is considered more than a family affair. The father of a British soldier killed in Afghanistan concluded his letter to a British newspaper as follows: 'Sorry these boys are dying out there, but at the end of the day who is going to remember them after they die?? Not the politicians, that's for sure' (Allen and Groves 2009).

The father's accusation was not entirely fair; as shown in the previous chapter, political leaders often participate in commemorations of the dead. And they are not the only ones. The armed forces themselves do most of the commemorative work, playing a highly visible part in the remembrance ceremonies of every country. In Denmark, when coffins are taken out of the airplane and brought into the reception hall, the officer in charge promises that the name of the dead will be honoured forever and that his name will be carved on the regiment's memorial (Skytt 2010). In Kabul, the German armed forces contingent set up its own, makeshift memorial soon after the first Bundeswehr soldiers were killed. On it the soldiers inscribed: *In honour of the dead* (*Zeit* 2007: photo 15). That was long before anything similar was erected in Germany itself. Britain's Remembrance Sunday, which commemorates all of the country's fallen, is a thoroughly militarised affair, whereas Germany's public commemorations are purposely civilian and so retrospective in nature that the soldiers who died in Afghanistan have long seemed an almost embarrassing disturbance.

The prominence of the armed forces in officially recognising the dead is easily explained, even in the special case of Germany. Ceremonies are important to active-duty soldiers and veterans because, as they often say, they feel a debt to their dead comrades. Senior officers note that official commemorations remind people at home that their country's men and women in uniform are willing to pay the ultimate price to protect them. Not surprisingly, the strongest advocates for official recognition in the form of remembrance days or monuments have been veterans' organisations and soldiers' associations. They are keenly aware of the linkage between official recognition of war casualties and their own status in society. Another reason has to do with professional identity. For the armed forces, commemoration is an important element of socialising the rank and file and enhancing their sense of belonging. The British Field Marshall Alexander once explained that public military ceremonies such as parades and wreath-laying are important because they let the public see soldiers expressing 'pride in their performance, pride in their Regiment or Corps and pride in the profession of arms' (Gordon 2002: 70).

Quite another aspect of such exercises, rarely mentioned but palpable nonetheless, is the exposure to death that they provide, permitting soldiers to confront and overcome their own vulnerability (Møller 2010: 225–32). This may be one reason for the impressive attendance at repatriation ceremonies held in Afghanistan before coffins are loaded onto aircraft for transport home. In the written accounts of soldiers, whenever such ceremonies are described the emphasis is on the full attendance: 'The Army's one big team at the end of the day' (Danziger 2010: 42). The extensive military participation in funerals and commemorations at home can also be seen as a delayed response. On the field of action there is rarely time to confront the sadness of losing a team member. Many written accounts from Afghanistan refer to the limited time available. As one Welsh Guards officer in Helmand put it: 'We will remember our mates but we're here do to the job first. … There'll be time for beers and tears when we get back' (Harnden 2011: 414–15).

For politicians, public ceremonies increase the visibility of the armed forces in society. With national conscription either abandoned, as in Denmark and Britain, or severely reduced as in the German case, fewer and fewer members of the general population have direct contact with the military. During the Cold War, when the armed forces were significantly larger and more focused on homeland defence, seeing men and women in uniform was far more common than today. Commemorations help fill the vacuum. According to a British Parliamentary Report, ceremonies may also boost the armed forces' standing in society and thus merit defence spending (Davies et al. 2008: 4). This aspect is particularly relevant during a war with as little public support as the one in Afghanistan.

For the next of kin, the benefits of ceremony are altogether different. In addition to providing public recognition of their loss, public commemoration helps protect against oblivion. When a Danish newspaper prepared a series of in-depth presentations of every man and woman killed in Afghanistan, only three of 38 families did not want to share their memories with the journalists (Stougaard et al. 2010: 19). The same positive response greeted the British artist Steve McQueen when he approached the families of dead soldiers for photos he wanted to use in an art installation where the faces of the dead were to be reproduced on postage stamps. Of 102 families, only four declined (Searle 2007). Those who held back gave the same reason in Britain and Denmark: the loss was so recent that any publicity would be too traumatic. But most families, felt that sharing the name and face of their loved one would help keep his or her memory alive, even if no direct comfort was to be obtained. All who sent in a photo approved of the resulting art project. In the words of one father: 'It's reminding people – when you buy a stamp, you see somebody's face' (*The Scotsman* 2008).[1] Interestingly, apart from the four families that declined to participate, the only negative response came from the Ministry of Defence. It claimed that images of the dead were too emotionally powerful to be exhibited in the middle of a war (James 2010). Most families felt no such inhibition.

The families, the armed forces and the politicians define how the dead are perceived, received and commemorated. The outcome in the form of memorial days and monuments is a valid indicator of society's attitude to war. A look back in time shows how much has changed. The clear linkage between patriotism and war, first made during the French revolutionary wars and with the introduction of national conscription, continued into the nineteenth century. Militarism and national expansion went hand in hand. The death of soldiers was idealised in monuments erected by cities, towns and villages all over Europe. The incredible loss of life in World War I gave rise to many forms of monumental expression. A common motif was that of the dead soldier shielded by an angel, endowing the fallen with an image of martyrdom on behalf of the nation (Mosse 1990). Other monuments were startlingly modern for the time. The London Cenotaph's simple, unornamented form broke from a tradition of overtly patriotic war memorials.

In Germany, commemorating the country's fallen soldiers after World War II was politically difficult (Koselleck 1994). Other victim groups had priority, and remembering them was understood as an integral part of coming to terms with national guilt (Eichmann 1985, Reichelt 1995). The difficulty is reflected in the monuments erected to commemorate all the victims of National Socialism; sometimes the names of dead Wehrmacht soldiers were added to the memorial tablets. The idea was to subsume the military dead into the collective of all those lost during the war, and to remember them all on the annual People's Day of Mourning in late autumn.

Memory Activists

A major difference among the three countries has been the role played by pressure groups seeking to have the war dead officially recognised. The groups consist of what's been called memory activists: men and women bonded by a common experience (Winter 2006: 136). They have had a clear agenda: to gain official recognition for those killed in Afghanistan. Their ability to succeed has depended on several factors. First and foremost is the extent to which they receive official support, and next is whether their agenda corresponds with the official framing of the war. Any gap between those who have served in the war and the politicians in charge makes it difficult to agree on a suitable form of commemoration. Conflicting views on the overall war aims are not the main problem. More vexing is the potential for disagreement over reality on the ground in Afghanistan.

1 Very soon, more than 20,000 signed a petition to have the photos made into stamps.

In Britain, there has been little such discord. Points of contention have included the lack of adequate military equipment and the failure to invest sufficiently in Afghan. But all these issues were debated in Parliament and the media, and none was sufficiently potent to derail the official version of the war as a necessary and justifiable response to al Qaida, or the need to honour the dead. In the official frame of view, the soldiers killed in Afghanistan dead are no different from the men and women in service who died in wars before them. Therefore, the main tenet of commemoration ceremonies has been continuity. The inscription on the Cenotaph – 'The Glorious Dead' – applies no less to the Afghanistan dead than to those who fell in Flanders almost a century ago.

In Denmark, by contrast, the early years of the Afghanistan war passed with no official commemoration of the dead – no memorial day, no monument. The Blue Berets, the Danish organisation for international operations veterans, has pressed since the Iraq War for some sort of public recognition of Danish fatalities sustained abroad. It cannot have been an arduous task, since the group's wish corresponded to the political agenda of the government. Danish officials have long understood the importance of a highly visible armed forces to make plain the country's change in security policy. When the Blue Berets proposed a memorial for those killed in international operations in 2002, the Minister of Defence included them in the planning process (Jensby 2002). As the war ground on, Denmark would establish both a memorial day and a monument, as we shall see.

Germany was a different case in that the pressure for recognition not only failed to correspond with the official version of the war, but actually contradicted it. The original German memory activists were the soldiers on active duty in Afghanistan who created their own makeshift memorial. Unlike the Danish soldiers, however, they played no coherent or visible role in efforts to have a memorial erected at home. That process, as will be shown, was conducted within the Ministry of Defence, and mainly behind closed doors. There are two veterans' associations in Germany. Their exclusion from planning for both a monument and a Veterans Day strained their relations with the Ministry of Defence (*Der Tagesspiegel* 2012). The other organisation that could have played a role in commemorating the country's dead was the German War Graves Commission. This did not happen. The Commission has steadfastly maintained a low profile whenever these issues have been on the agenda. It remains focused on the maintenance of Second World War graves, and Afghanistan is rarely if ever mentioned in its official publications (Wernstedt 2009). The Commission's emphasis on reconciliation, originally intended to heal the wounds after German atrocities in occupied Europe, was less transferable to the Afghan war.

Memorial Days

Memorial days are official nationwide dates when a specific historical event is remembered. Often the original cause is forgotten or barely noticed apart from the explanatory line on red-letter dates in calendars and almanacs. Although participation in the early years may be massive, memorial days and monuments can fade from collective memory with surprising speed. In most Danish cities, towns and villages, a commemorative stone or statue marks the return of Southern Jutland to Denmark after Germany's defeat in World War I. For decades afterward, the monuments would be decorated with flowers and wreaths on the official Reunion Day, 15 June. Gradually the annual event lost its potency, and the monuments are now little more than historical relics. Likewise, 18 April, which became an official day of support for the Danish minority in Germany soon after Denmark's reunion with Southern Jutland, is no longer celebrated officially. Lapel pins once sold in support of the Danish minority and later worn as a sign of patriotism have fallen from usage, and sales were stopped in 2008 (Grænseforeningen 2010).

The French historian Pierre Nora has documented how days like 18 April lose relevance. According to him, commemorations are fuelled more by memory than history (Nora 1989: 8). When the government unveils a monument or establishes a day of national commemoration, public perceptions are influenced only for a short time, often just a matter of years. The monuments remain but mean little to passers-by, and the date ends up only as a red-letter day in the calendar with a few words underneath noting its origin, as is the case with 18 April. If ordinary people appropriate a monument or a special day for themselves, perhaps in ways not foreseen when it was established, then it relevance is assured. When the Vietnam Veterans Memorial in Washington was being planned in the early 1980s, few foresaw how warmly its Memorial Wall would be received by veterans and by the families of the fallen whose names are inscribed on it. Visitors deposit letters, flowers, and mementos at the recessed wall as they find the names of those they lost. Nearby is a far more traditional, heroic-style monument to the war, but it has never attracted the same amount of attention and participation. The political potency of the Memorial has remained considerable. In 1983, Ronald Reagan, a staunch defender of the righteousness of the Vietnam War, visited it (Bates 2011: 66–86). At the time, popular interpretations of the war itself remained unsettled, with more focus on why the war had been fought than on those who had fought it. Reagan's visit was an act of reconciliation.

Germany: Shunning the Dead

The People's Day of Mourning, celebrated since the interwar years, is the annual occasion for commemorating Germany's war dead. The main ceremony takes place in the Bundestag in Berlin, with the Federal President saying a memorial prayer. The keynote address is usually given by members of the governing cabinet or the opposition, though in 2008 the Prime Minister of Luxemburg, Jean-Claude Juncker, was invited to give it. The leader of the German War Graves Commission also speaks. Compared with British commemoration ceremonies, it is a sombre occasion. There are no uplifting military marches or folk tunes interspersing the sorrowful music played.

The Day is a good indicator of how the Federal Republic has officially come round to acknowledging the Afghanistan war dead. An examination of speech transcripts shows that long after 2001, when the deployment began, speakers retained their traditional retrospective historical orientation. Despite mounting German casualties in Afghanistan, attention was given exclusively to the victims of the two world wars and to Germany's role as the perpetrator of genocide. Current events were brought up only to remind the audience that the world was not at peace; that men and women were still dying in conflicts and wars despite the horrible lessons that could be drawn from the German experience. Not until 2005 were Bundeswehr casualties in Afghanistan given attention. Minister of Defence Peter Struck ended his speech that year by asserting that he bore the full responsibility for German soldiers in Afghanistan, and that as a consequence each death touched him deeply (Struck 2005). A half-year earlier, as mentioned previously, Struck had been ridiculed for stating in parliament that Germany's security was being defended at the Hindu Kush (Struck 2004). In the years following his speech, the Bundeswehr deaths suffered in Afghanistan would be mentioned during the ceremony in the Bundestag with one major difference. While Struck had openly attributed German casualties to attacks, the references to Afghanistan in subsequent years specified the mission's character as a peace operation, and any mention of how the soldiers were dying was avoided. In 2010, for example, Foreign Minister Guido Westerwelle paid tribute to those in service while at the same time enhancing the notion that Germany is different from other countries: 'Germany will also in the future stand for a culture of restraint concerning the use of military power. The deployment in Afghanistan shows us how dangerous the work for peace and human rights is' (Westerwelle 2010). Shortly before Westervelle's speech, Federal President Christian Wulff had lumped Germany's recent military dead with many other victims of violence: 'We mourn the victims of today's wars and civil wars, the victims of terrorism and political persecution, the Bundeswehr soldiers and other forces deployed who have lost their lives abroad' (Wulff 2010).

Compared with Struck's speech, those of Westerwelle and Wulff seem timid, reflecting a strong reluctance to depart from the retrospective frame given to the People's Day of Mourning in the Bundestag. Such reluctance is closely tied to the Federal Republic's political identity as utterly different from that of the Third Reich (Sabrow 2009). Even passing references to dead Bundeswehr soldiers have been interpreted as a dangerous diversion from what the day is meant to be (Echternkamp 2008, Hettling 2008).

With the People's Day of Mourning unchanged, it is relevant to ask whether there have been other occasions when Germany's dead soldiers have been acknowledged to a wider audience. The funerals in 2009 and 2010 were unique occasions, and unlikely to serve as a model for the future. An annual event that could have lent itself to an official recognition of the dead is the annual Ceremonial Oath by new recruits in front of Parliament.

Until the mid-1990s, this affair was open only to family and friends and conducted on military ground. In 1996, Minister of Defence Volker Rühe started what was called a 'pledge offensive'. The idea was that by making the ceremony public, the relationship between the armed forces and society would grow closer. After long deliberation, it was finally decided in 2008 to transfer the ceremony to an open ground in front of the Bundestag. The question of official attendance remained unsettled until, at the last minute, Chancellor Merkel and her Defence Minister declared that they would attend (Heckmann 2008). As it turned out, they were among very few politicians to do so. A former General Inspector of the Army, the highest ranking military officer in the Bundeswehr, called the weak attendance an affront to the men and women risking their lives (*Handelsblatt* 2008). It rendered hollow, he said, the often heard reference to the Bundeswehr as a 'parliamentary army' (Mayntz 2008).

His criticism and the media's attention may explain why Merkel delivered the official speech at the following year's Ceremonial Oath. By then, German soldiers had fallen in Afghanistan, and as she spoke they were engaged in a large combat operation against the Taliban. The media had, with unusual frankness, described it as very dangerous (Gebauer and Najafizada 2009). It seems fair to assume that the fight was on the mind of both the recruits and the families present. Merkel (2009) included Afghanistan in her speech, lauding those serving under 'difficult conditions and high risks'. Apart from that brief reference, she chose instead to focus on Colonel Stauffenberg, the German officer who plotted against Hitler and was executed, and on her own experience growing up in Communist East Germany, deprived of the freedoms that the recruits were pledging to defend. The democratic heritage of the Bundeswehr, in particular its difference from the East German armed forces and Hitler's Wehrmacht, was her underlying theme. The oath ceremony, like the People's Day of Mourning, stayed retrospective. In 2010, a former concentration camp prisoner gave the speech, and in 2011 the Federal President spoke without mentioning Afghanistan or the seven Bundeswehr soldiers who had been killed so far that year (Wulff 2011). In 2012, the ceremony was transferred to the Ministry of Defence for reasons that remain unclear. It is unlikely that public criticism was the cause. It is true that some peace campaigners and politicians had opposed converting the oath ritual into a public ceremony, but they were few. The effect of moving it to the ministerial grounds has been to eliminate the only event at which the public could see the armed forces parading and gain a visual impression of the relationship between political leaders and the military. The Ministry of Defence is not easily accessible and its surroundings are far gloomier than the open space in front of the Reichstag. The number of onlookers in 2012 was far fewer than in 2010.

Both the Ceremonial Oath and the Bundestag mourning ceremony could have served as occasions for Germany's Afghanistan dead to receive homage. The failure to use the ceremonies in that way is a source of discontent among the veterans (Dirk 2009). To lessen tension, Minister of Defence Thomas de Maizère has proposed making the People's Day of Mourning into a Veterans Day. Although the Minister claimed inspiration from Veterans Day in the United States, similar ideas have been discussed among Bundeswehr officers for quite some time, and they surfaced in the media in mid-2010 (Wenzel 2010). The German Federal Armed Forces Association immediately came out in support of the proposal when it was made public (Sebastian 2012). The only elected officials to react were those on the left, who rejected the idea immediately. A few balked at the date favoured by the defence minister (*Spiegel* 2012). If the Day of Mourning were to be transformed into a day for honouring veterans and dead soldiers, critics said, it would all too easily resemble Hitler's Heroes Memorial Day. It was therefore a thinly wrapped attempt to reintroduce hero-worship, whose consequence would be 'a remilitarisation of society' and an end to Germany as 'model to the rest of the world as a civilian power' (*Süddeutsche Zeitung* 2012).

At the same time, signs were emerging that lower levels of the political system felt less reluctance to acknowledge German deaths in Afghanistan. In towns and villages where the dead were buried, participation by local politicians was customary. The mayor of a Bavarian town decided in 2011 to add the names of three local men killed in Afghanistan to the local war memorial. He wanted 'to develop a culture of participation' (Jungholt 2011). As far as can be ascertained, it was the first time the names of dead Bundeswehr soldiers killed in combat have been inscribed on a public monument.

Denmark: Creating Tradition

In 2009, the Danish parliament decided to make 5 September the official 'Day for Danish Soldiers Deployed Abroad'. The cumbersome name was soon forgotten, with 'soldiers' memorial day' entering common usage. The popular name also better reflects what the day has come to mean in the short time since the first celebration. Apart from a military parade in the capital and the other larger Danish cities, the main event is a memorial service with the royal family and the government present.

The proposal for a memorial day was made in 2006 by a soldier serving in Afghanistan. Impressed and inspired by the way British and American memorial ceremonies at home enabled the public to express its support for the troops, the Danish soldier wanted something similar to be arranged in Denmark. He had seen that the ceremonies mattered not only to families at home but also to soldiers in Afghanistan 'when the artillery shells rained down' (Hansen 2009a). He set up a website to gather support and media attention. He was successful on both accounts. However, the initial response from the Ministry of Defence was negative because the number of national flag days was already too large to allow for yet another (Nielsen 2008). However, the attitude soon changed. One reason may have been the appointment of Søren Gade as minister, with his famously pro-soldier attitude. In addition, the soldier's website generated thousands of hits and support messages in the course of its first year, attracting the attention of journalists and politicians.

The main point of discussion became not whether the day should be introduced, but when. The political parties were well aware that the choice of date would resonate with historical significance and potential controversy. The officer who launched the idea wanted to use 5 May, Liberation Day, which commemorates the end of the German occupation in 1945. The soldiers deployed in Afghanistan would then share a lineage with World War II freedom fighters. The date had the added benefit of being a flag day already, so a new one would not be needed. However, no one of prominence came out in support of 5 May. Instead, the government proposed 31 September – the date in 1990 when the Danish parliament had agreed to send a naval vessel as part of the US-led coalition to liberate Kuwait. As mentioned previously, that deployment marked the start of the transformation of Danish foreign and security policy away from home defence and toward international engagement. The government would probably have had its way if the Progress Party, which usually supported the government on defence issues, did not prefer 29 August. On that date in 1943, the Danish government resigned in protest over the growing interference in domestic matters by the German occupants. Nobody within the Progress Party was able to explain the possible connection between 1943 and Denmark's current deployments, and the proposal fell. The government was left with no clear choice. After what must have been a remarkably short period of deliberations, it settled on 5 September. This day had no historical connotations. The government's haste may have been due to public embarrassment over the juggling of dates (*Jylands-Posten* 2007, *Information* 2008). In any case, having an official celebration of the troops was very much consistent with official policy. Moreover, to judge from the popular response and the impressive catalogue of arrangements that has since arisen on the date, many Danes had wanted some way to express their support for the troops.

Lobbying by members of the armed forces had played an important role in creating the special day. When 5 September arrived, memorial ceremonies were held all over the country, with military parades, wreath-laying and religious services (Folk & Forsvar 2009). The main celebrations were in Copenhagen, with attendance by the Queen and the government. The Prime Minister gave the memorial address during the service. His speech set a pattern that has been followed, focusing narrowly on the laudable decision of soldiers to serve in Afghanistan (Rasmussen 2009, 2010, 2011a). They were presented, again, as the bearers of Denmark's finest national qualities. The importance of their deployment in the protection of Denmark against terrorist attacks was repeated. How entrenched this framing had become was evident in 2012 when the new Prime Minister, this time a Social Democrat, gave the memorial address (Schmidt 2012). No change in tone or emphasis could be detected.

The Afghanistan war itself, including the prospects for success and the government's responsibility for soldier safety as a consequence of deploying them, has never been mentioned. In Germany, by contrast, the Bundestag's decision to deploy the military is often emphasised, and when the Chancellor and Minister of Defence talk about Afghanistan they inevitably note that the soldiers are there on behalf of all Germans. In Denmark, the strong emphasis on individual choice has deflected attention from the issue of consensus. Officials have sought to interpret the public response to memorial arrangements as support for government policies, but the officer who launched the memorial-day idea in the first place was very explicit in arguing that the day should be a chance for the public to express its support for the men and women in service (Stougaard and Caruso 2009). He had been inspired by the British soldiers he worked with in Afghanistan in 2006. The British media had managed to forge strong links between British soldiers in Afghanistan and the people at home, and he could see how important the backing was for morale. By way of example he noted that weather forecasts on British television included Afghanistan (Stougaard and Caruso 2009). Danish soldiers had none of this. His reflections on media interest tallied with that comment by the Danish editor: the war was no longer front-page material (Brøndum 2008).

When the first soldiers' memorial day was celebrated, in 2009, the dead were commemorated with religious services and with wreaths deposited at military monuments. The day ended with a military parade in the capital and some other cities. According to Danish media, popular participation was strong. The day met with little opposition. A few peace activists and politicians on the extreme left declared they would fly the flag at half mast (Hansen 2009b). A security scholar claimed that the day lacked a clear message and therefore democratic relevance (Sørensen 2010). It ran the danger of becoming a day only for the armed forces and the families of the men and women deployed.

This possibility cannot be entirely discarded. Yet it should be recalled that days with a predefined memorial intent, like Denmark's own 18 April commemoration, are easily forgotten when those with any personal memory of the celebrated events have died out. That is unlikely to be the case with 5 September. The 2009 event was for obvious reasons seen as honouring the dead from Afghanistan as well as the men and women still there. Yet, the date was deliberately chosen so that over time the connection with Afghanistan would loosen. The officer who launched the idea had originally favoured 23 July, the date of the first Danish injury in Afghanistan (Nielsen 2008). If he had prevailed, the special day would have been so closely associated with Afghanistan that the achievements of future deployments might go unremarked, especially those of a traditional, peace-keeping kind.

Britain: Consensus

In Britain, the commemoration ceremonies celebrated on Armistice Day (11 November) and Remembrance Sunday (which falls on the second Sunday in that same month) are nationwide events. On Armistice Day, two minutes of silence are observed at 11 a.m. all over the country in remembrance of the end of World War I. Remembrance Sunday is a far grander event, with military parades, speeches and wreaths deposited at the foot of war memorials everywhere. The procession in London down Whitehall along with the Queen's placement of the first wreath at the foot of the Cenotaph – an act then repeated by other members of the royal family, the Prime Minister and political leaders – are televised live every year. The number of spectators lining the streets is huge, and according to media reports it has been growing in recent years (Kennedy 2009). Most participants pin a red paper poppy to their lapel. In the weeks leading up to Remembrance Day, the pins are sold by the Red Poppy Appeal, a charity working for war veterans. The wreaths, too, are made up of these poppies. The tradition dates back to a poem written by John McCrae, a Canadian officer, in 1915:

> In Flanders fields the poppies blow
> between the crosses row on row.

The poem became widely read and was soon associated with the soldiers dying on the Western Front.

The poppies express the strong sense of history and continuity that characterises British commemoration ceremonies. The changes that have occurred have been subtle, adjusting to collective memory as time passes. In the interwar years, the emphasis was clearly on the sense of loss suffered by almost all. On 15 November 1920, *The Times* compared the enormous stream of visitors to the Cenotaph and to the Tomb of the Unknown Warrior in Westminster Abbey to a Lourdes pilgrimage (King 1998: 21). Men would doff their hats when passing the Cenotaph, a gesture no longer performed. The mourners are no longer in majority; instead attention is focused on the Queen and the royal family, the government and opposition leaders deposing their wreaths.[2] This was how the first commemoration ceremony had been intended in 1919 (Lloyd 1998: 50–52). The political and military elites were to have played the main role in what was to have been a celebration of Britain's victory in the Great War as well as an observance marking the burial of an unknown soldier. Instead, popular pressure transformed the ceremony into an occasion for publicly grieving the loss of all the dead. Lord Curzon, who was responsible for war commemorations, understood this, and when the first wood-and-plaster monument was replaced with the white stone structure we see today, he issued instruction that priority should be given not to 'society ladies or wives of dignitaries, but to the selected mothers of those who have fallen, especially in the humbler ranks' (Calder 2004: 16). With fewer mourners, the day has become both a day for remembering the dead and for celebrating the country's military traditions.

Another change can be seen in the use of the poem *For the Fallen*, written by Laurence Binyon in 1915. The lines 'They shall not grow old' and 'We will remember them' are carved on many war memorials and printed on the ribbons decorating the poppy wreaths. Especially the fourth of the seven verses have remained popular; commonly known as 'The Ode', and is often read out during commemorative services:

> They shall grow not old, as we that are left grow old:
> Age shall not weary them, nor the years condemn.
> At the going down of the sun and in the morning,
> We will remember them.

2 And if a slight breach of protocol occurs, comments in the press are ruthless (Meltzer 2009).

At the end of reading, audiences of the past would traditionally repeat the final line. This practice has largely disappeared and with it the scope of the commemorations has widened to include both sorrow over the dead and military pageantry (McIntyre 1994: 73). This mixture is reflected in the music performed, from the patriotic '*God Save the Queen*' and '*Rule Britannia*' to sad dirges like Beethoven's '*Funeral March*', Elgar's '*Nimrod Variations*' and the bagpipe lament '*Flowers of the Forest*'. The differences in style convey the complex purpose of the day as an occasion for mourning the dead and taking pride in the country's military traditions.

The pride and glorification of the armed forces would likely dominate the day were it not for the highly visual role played by war invalids of all ages and from all recent wars and operations. The sight of the mutilated young in particular, often struggling on crutches to deposit wreaths for their dead comrades, gives onlookers a true sense of the costs of war. It serves as a counterbalance to the military pageantry of endless rows of marching men and women in uniform, accompanied by their own regimental bands. Alex King, in his analysis of World War I memorials, has shown that members of planning committees at that time often wanted the living to draw strength from the dead (King 1998: 216–45). Today's mix of soldiers in current service with veterans and marching bands can be seen as a fulfilment of that ambition.

The absence of any concerted criticism can be taken as a proof that most British citizens agree on how to honour and commemorate the dead. Popular participation is astounding, both during the two minutes of silence on Armistice Day and during Remembrance Sunday, when red poppies are seen everywhere. When Jon Snow, a Channel Four news presenter, refused to wear one as he read the news, he attracted considerable criticism. Snow then accused his opponents of 'poppy fascism' (Furness 2012). He was reacting to what he saw as the increasing commercialisation of remembrance. On this point, he received support from Ted Harrison, an installation artist who had been praised for his poppy installation in St. Paul's Cathedral. Harrison also objected to the way all wars and conflicts were treated as equivalent (Harrison 2012). While World War II had a clear moral justification, he said, subsequent conflicts have not, so they should not be ranked at the same level. The same argument had been made long ago, with considerably more force, by the theologian Ronald Coppin. In 1965 he wrote an article describing memorial ceremonies as glorifying war and elevating the dead into martyrs (Coppin 1965). Noting that people with a personal recollection of World War I would eventually die out, he added that 'we can only then both expect and welcome the gradual disappearance of Remembrance Sunday …' (Coppin 1965: 530).

Like Coppin before him, Harrison's objections failed to elicit much response, even though he had attacked the very premise of the British commemorations: that all the dead are equal, irrespective of the wars they fought. Shifting British governments have upheld the message. In his speech on Remembrance Sunday in 2009, Prime Minister Gordon Brown linked the recent victims in Afghanistan with those killed on the Western Front:

> This year's events will be particularly poignant because this summer we witnessed the passing of the last Tommy, Harry Patch. With his death, an entire generation has fallen silent and we have lost our living link with the momentous events which did so much to shape our nation. And so this Armistice Day, we must make anew our promise to Harry and his comrades: that although they are gone, we will never, ever forget. (Brown 2009)

When Brown was Chancellor of the Exchequer in 2006, he made an attempt to introduce a day to honour veterans. Although they were in focus on Remembrance Sunday, most of the attention went to the dead and not the survivors. According to a committee appointed by Parliament, the public lacked a day in which to express 'that strong and latent feeling of appreciation and admiration which so evidently exists' (Davies et al. 2008: 4). An attempt had been made to hold a Veterans Day, but as the committee concluded in its final report, the day went by almost unnoticed by the media, and more importantly by the veterans themselves. What was needed was a day with much greater media appeal. A petition had been organised to turn the Monday following Remembrance Sunday into a 'national remembrance public holiday to commemorate the fallen and our nation' (Honeywell 2008). The petition did not lack public support; MPs and senior officers backed it, as did the British Legion. In a little over a half-year, more than 500,000 people signed the petition. The petitioners were heard, and the committee proposed a Veterans Day. The only objection concerned the date. Remembrance Sunday had become 'a familiar and cherished institution' (Davies et al. 2008: 18). Veterans Day was therefore assigned to the final Saturday in June, and the first observance took place in 2006. In 2009 it was renamed Armed Forces Day. Judging from the press reports, honouring the fallen has been an integrated part of the day all over the country (Barrett 2009, BBC 2009, *Daily Mail* 2009, Prince 2009). The name change had no effect on that.

Internet Memorials

No monument devoted exclusively to those killed in Afghanistan has been erected in any European country so far. The monuments discussed above all antedate the Afghanistan war. However, a different and new form of memorial specifically commemorating the Afghanistan dead has emerged on the Internet. To date, most of the websites are the work of fellow-soldiers or friends, and most are devoted to one person only (BT 2008). How long they will be maintained is uncertain. Those without official backing, meaning the vast majority, will gradually turn into historical mementos – something like the monuments to forgotten generals and battles found in most European countries. Only in Britain and Denmark have memorial pages to the fallen been launched on official websites. The UK's Ministry of Defence site devotes a page to each soldier lost (MoD 2011). In Denmark, the task has been less centralised, with pages created by the regiments and posted on their own websites. At the start of the war not everyone was honoured in this way, but as the end has approached the dead soldiers have all had their commanding officer's eulogy reprinted. All German memorial sites have been privately launched. With no official websites available, fellow soldiers and next of kin have made their own.

The British MoD memorial web pages share some basic features with the lists of fallen soldiers published officially during previous wars. For example, each soldier's name, rank, age and place of death are mentioned. What is new is a lengthy description of each soldier's personal qualities, both as a team member and soldier: 'He was a fun, friendly, talkative character who always wanted the best from his dog, his troops, and himself' (MoD 2011). As always, the pledge is made that he 'will always be remembered'. The solemnity of these words stands out in what is otherwise a low-key, personalised tribute. Emotions tend to be stronger in the brief messages left by girlfriends, family members and fellow soldiers. The addition of their lines beneath the officially endorsed profile and photos makes the ministerial website a meeting place between the ministry as the authority ultimately responsible for sending soldiers off to Afghanistan and the grieving public. It also helps those who knew the dead to make their loss known and share their grief.

Danish memorial sites contain the bare biographical information of those killed, followed by tributes paid by their commanding officers. Yet there is little of the objective, factual tone found on the British site. The Taliban fighters responsible for Danish deaths are described as cowards, the lethal explosions as 'merciless'. The Danes who died are said to have been fully aware they might end up 'paying the ultimate price', and the circumstances of death are described. Instead of closing with a promise to always remember the deceased soldier, the officer in charge declares that the death 'has not been in vain' (Hæren 2009). The soldier's combat abilities are praised, and often his dedication to the military profession is underlined (Hæren 2008). After the lengthy profile, colleagues add reminiscences and promise not to forget.

The anonymity of being one name among many is shattered on these memorial sites. Web technology makes it easy for friends and family to leave final words to the dead, but there is nothing new about the impulse to do so. Such communion occurs at some physical monuments, too, even if the designers did not plan for it. The Vietnam Wall is well-known for this. On the memorial websites, a sense of loss pervades all of the privately composed entries, whose writers seem purposely to avoid remarks that could disturb the melancholy atmosphere. Even on private memorial sites, where one might expect less intensive monitoring of entries, I failed to find any comments that differed in tone or content from those posted on the official sites. Likewise, none of the posters attempted to defend, let alone glorify, the war. On the memorial web page for a 24-year-old Danish officer killed in Helmand province in March 2008, the privately added lines include: 'You tried to make a difference, but paid the ultimate price. Honoured be your memory!' and 'I think of you. Take care, wherever you are' (Garderportal 2008). Almost identical lines have been posted on German and British websites.

Perhaps the most striking feature of these Internet memorials is individualisation: the fallen are referred to by name and rank. An honour previously bestowed only on acclaimed heroes or long-serving generals warranting their own statue is now democratically available to every dead soldier whose loved ones wish to share their loss.

Monuments

The most common monument to the dead is a tombstone or a cross. It is a private, individual memorial, meaningful only to those that knew the one whose name is inscribed. When placed together in long rows covering vast areas, as on some World War I battlefields, the vast number and the monotony of the rows create a powerful totality. The effect is similar, when the multitudes of red poppy wreaths, are placed at the foot of British war monuments on Remembrance Sunday each year.

A war memorial is different. There are statues and plaques to military leaders and particularly brave soldiers, but most memorials are about collective sacrifice, as expressed in the single line carved on the London Cenotaph: 'The Glorious Dead'. The individual soldier disappears into a greater community of war dead (Münkler 2007). At the same time, a war memorial is an expression of political values, with patriotism as a common denominator in monuments erected before World War I. In those unveiled after 1918, the loss suffered by the local community or the country play a more prominent role in the design, although patriotic symbols and inscriptions are by no means unusual.

Jay Winter (1995), who has written extensively on World War I commemoration, warns against overemphasising the political contents of these monuments at the expense of what he sees as their existential role as receptacles of grief and mourning. Once those personally affected by a great event die out and collective memory fades, public mourning at the commemorative site becomes more of a formality (Halbwachs 1997: 97–142). That is not to say monuments inevitably lose relevance; their significance can be reinterpreted, giving them a new lease on life. But if a monument is to retain its original potency, and be more than just another picnic spot, it must remain clear to those passing by or picnicking around it why it was placed there. In the words of the American sociologist Wendy Griswold, it has to be a 'shared significance embodied in form' (Griswold 1987: 13). The question is: shared by whom? Those who erected the monument may have had a specific understanding, not known by all, of what they wanted the monument to convey. It is not just a question of artistic form or inscriptions. In the case of a war memorial, it matters that the framing of the war expressed in the monument is shared by the rest of society. Expressing consensus is easier in memorials that commemorate a particular battle than in those dedicated to the fallen. Honouring dead soldiers is more difficult because they make the war personal; their suffering and sacrifice deepen the emotional impact on the viewer. When a war loses public support as the years pass, it can become difficult to find Griswold's 'shared significance'. It was inevitable that the framing, and in Germany's case controversies over the Afghanistan war would influence decision-making and public opinion about memorialising those who died in it.

Some monuments seem intentionally designed to avoid any attempt at expressing a relationship to any particular war or dead soldiers. One example is a red bear carved in stone that has been donated by the city of Berlin and erected at Kabul Airport (Rathaus aktuell 2003). On the Bundeswehr homepage, it is listed as a memorial to the Bundeswehr units that have served in the country. Why is unclear, because there is nothing commemorative about its appearance. As the city symbol of Berlin, the bear has no evident connection to the war or to Germany as a whole. Moreover, no ceremonies are performed at the memorial; it seems as if it is not used in any way. If that continues to be the case, the bear, like any other disused monument, will be forgotten. Even in Britain, where commemoration of the war dead remains an important part of official life, war memorials placed out of the public eye often end up as derelict, vandalised ruins (Duffin 2012).

In their analysis of the Washington Vietnam Veterans Memorial and its unique wall of names, Robin Wagner-Pacifici and Barry Schwartz ask: 'When is a war memorial no longer a war memorial?' (Wagner-Pacific and Schwartz 1991: 381). Answering their own question, they note that memorials over time can be sapped of their original meaning. If the statue of a heroic soldier ends up, for instance, as a picnic spot or a skateboarders' hangout, it may be reduced a decorative backdrop. Alan Borg's presentation of British war memorials contains an illustration of exactly how radical such a change can be. The impressive Grecian Hall of Remembrance in Stockport, which commemorates the fallen in the two world wars, is photographed with a banner stretched across the portico welcoming visitors to 'A Celebration of Cartoons and Comics' (Borg 1991: photo 204).

If veterans of a war gather on a special day to remember their dead comrades, casual onlookers and passers-by realise that some of their countrymen remain emotionally moved by the deaths; those outside the circle of mourning may give fresh thought to the war and those who perished. This has been one of the important lessons of the Vietnam Memorial Wall, where people gather not only to see the lists of names carved but to observe other visitors touching the names in recognition of a loved one or comrade (Wagner-Pacific and Schwartz 1991: 403). For many veterans, such visibility is important; to them, a monument is not just a place to gather in tribute but a reminder to the rest of society that some wars and peacekeeping operations are unjustly forgotten. By showing others they have served, their contributions are not hidden but publicly recognised. An example of this is the presence of war veterans during Britain's Remembrance Sunday in November.

Sometimes a reinterpretation of the past leads to the renewal of an existing monument or the creation of new ones recalling a war long past: The Monument to the Women of World War II, unveiled near the Cenotaph in 2005, is a recent example of the latter, as is the Shot at Dawn Memorial from 2001, recognising deserters from World War I. The Shot at Dawn motif would have been unthinkable in the decades after the 1918 Armistice, when deserters were looked upon with none of today's understanding for their actions. An example of how a monument has recently been endowed with a message opposite of the original intentions is the Danish Isted Lion, originally unveiled in 1862 at a cemetery in the then-Danish city of Flensborg to commemorate a battle fought against Prussia in 1850. Two years later, after Denmark had lost against the same opponent and Flensborg became Flensburg, the Lion was moved to Berlin as a war trophy. In 1945, the statue was discovered at an abandoned military academy in Berlin. The fate of the Lion had 'caused a resentment which till this very day is very much alive in wide circles of the Danish people', according to the Danish foreign minister at the time (Kurtz 2006: 134). The Lion was promptly returned to Denmark and placed in Copenhagen, only to be returned again to Flensburg (still a German city) in 2011. The Lion, originally regarded by the city's German minority as unabashedly nationalistic, was now framed as a symbol of reconciliation by the Danish prince presiding at the hand-over ceremony and by the German mayor receiving it (Lohse 2011).

The Danish–Prussian war is no longer, of course, a subject of controversy. When historical narratives are actively in conflict, erecting a monument may be impossible. In Britain, that was the case with servicemen killed in Northern Ireland by the IRA. Discussions over a suitable form of commemoration have continued for quite some time without conclusion (Rayment 2007). In Germany until very recently, the only monuments to dead Bundeswehr soldiers where at military bases, off-limits to the public. Constructing a public memorial had been judged too provocative. Glorifying dead soldiers had been an integral part of National Socialist ideology and was therefore politically taboo. When it finally was decided to create a memorial to all who have died in service, including those killed in combat, and to place it in view of everyone, debate flared. Opponents claimed that a monument would encourage hero-worship and idealisation of the military. It was a singularly German controversy. Yet the basic question of how to honour fallen soldiers in a way corresponding with national political ideals has had to be confronted in Denmark and Britain as well. The debate over monuments is revealing because it shows the extent of social consensus with regard to the justness of the war in question; it also shows the nature of public regard for dead soldiers and veterans (whether as heroes or as incarnations of an outdated masculinity) and the likelihood of successfully integrating recent history with the collective memory of wars past. Commemorations are not fixed phenomena, no matter how permanent the stone of a memorial. Some are recreated and some fade. Ken. S. Inglis, who has written extensively on war memorials, concludes that most structures have grown 'too familiar for us to see them as History' (1992: 5). Some observers might consider that judgement an unintended compliment to the artists and political leaders behind many such memorials. Those familiar enough to transcend history, instead of being left behind, are likely to have reflected publicly relevant interpretations over time and thereby remained touchstones of grief and pride for generations of war veterans and families. Such acceptance has not been achieved with equal success in the three countries examined here.

The German 'Avoidance Monument'

The most controversial of the three national monuments is without doubt the Berlin Honorary Memorial to the Bundeswehr Dead, inaugurated in 2009. It is located on the grounds of the Ministry of Defence complex in Berlin. It stands at one end of the Paradeplatz, or parade ground, facing the main entrance to the ministerial building.

The memorial is a narrow rectangular structure stretching more than 30 metres. The shorter walls are of concrete. The long bronze facades are imprinted with large ovals, some split in two, meant to recall the identification tags worn around the neck by all soldiers during deployment. When a soldier is killed, half the tag is broken off and taken back to base as proof of death. The symbolic significance of the monument's ID-tag theme becomes clearer if we imagine replacing the tags with swords, so often used on war memorials erected before and after World War I to symbolise the end of battle. The Berlin monument contains no military symbol apart from those stylised ID tags. As ornamentation, they immediately drew criticism, with some observers detecting an attempt to endow soldiers who gave their lives with pre-eminence over other victims (Mittig 2007). It should instead be seen as expressing the collective identity of the dead as members of the Bundeswehr. With the dog-tag motif, no distinctions are made among the dead; they all died on duty, as soldiers. The monument may thus be seen as part of what the German historian Reinhart Koselleck has called 'the democratisation of death' (Koselleck 1979: 269).

Democratisation is also evident in the inscription: 'To the dead of our Bundeswehr, for freedom, justice and peace'. The honourees have fallen in defence of the political ideals of the Federal Republic, not for Germany. The structure embraces all who who died in Bundeswehr service, regardless of mission or cause of death – a clear signal that it's no warrior memorial. However, the timing of its construction has made it a monument to the Afghan effort more than anything else. The politicians in charge have done much to secure that identification. In the early summer of 2007, Defence Minister Jung seized on the death of three soldiers in Afghanistan to declare that it was high time a memorial was erected (Jung 2007). He said the need had become clear to him during a visit to the German camp in Afghanistan two years earlier, when he came upon the makeshift memorial constructed by the men serving there (Speicher 2009).

The Berlin monument may be entered. The inside is a large 'monochrome, dark, dematerialised chamber where the physical limits of the room seem to dissolve' (bmvg 2009: 25). It gives the visitor a chance to stand in silence and mourn the dead. Some of those opposed to the idea of having a public monument to lost Bundeswehr soldiers have portrayed this room as an attempt to create a religious-mysterious atmosphere where the dead are transformed into heroes (Brendle 2007).

If there is any such effect, it certainly shatters as visitors turn to leave. At that moment, their attention veers to the LED lights on a concrete beam above the exit, spelling out the names of the dead. Each name appears for about five seconds. The beholder is left to wonder about each soldier's age and rank and the year and location of death. The approximately 3,200 names appear in alphabetical order. Were the names grouped by year of death, each death would not have seemed like an isolated event but rather the consequence of a specific operation or deployment. As shown in the preceding chapter, the sense of belonging is very strong to soldiers. The jumbling of names deprives each one of context.

A relevant comparison can be made between this inner room, or cella to use the official term, and a very similar one inside the New Guard House [Neue Wache] on Berlin's main boulevard, Unter den Linden. Today, the building is The National Memorial to Victims of Tyranny and War. Built in 1816 in a strict neo-classical style with a portico of Doric columns, the open space inside is dominated by a statue of a grieving mother holding the body of her son. It takes some time before the visitor adjusts to the changes in light and takes in the statue, situated under a small, open skylight. By contrast, in the inner room of the Bundeswehr monument there is no time to pause and adjust as you try to read the names of the dead appearing one after the other.

Part of why the LED lights seem so glaringly at odds with the atmosphere of the room is the fact that they were a late addition to the original design, for which names had been specifically rejected (FAZ 2007). As first intended, the monument would have adhered to the official informational practice whenever a Bundeswehr member is killed, with no details beyond the brief communiqué disclosing the loss of life. When debate about the naming emerged in the media, *Stern*, a widely read newsmagazine, wanted to publish a series of portraits of the dead with photos and a description of where and how they had been killed (Jörges 2007). The Ministry refused to assist in any way. Eventually the journalists discovered that they could get all the information they needed from US sources. The ministry's justification for withholding the names from the memorial was that it was up to the next of kin to decide. It soon became evident that this view had little support among soldiers and the afflicted families. In the end, the minister relented and the names were added. By that time, lists with the names compiled by private persons had already been accessible on the Internet. Finally, in the summer of 2011, the Bundeswehr posted online an official list containing the names and years of death (Bundeswehr 2011).

The initial withholding of names from the memorial was not the only topic of debate. Still more controversial was the choice of location. The official rationale for the winning location was its historical connotation. In the ministerial courtyard, Colonel Claus von Stauffenberg and two of his accomplices were executed after the failed plot against Hitler in 1944. The Bundeswehr has on other occasions claimed the plotters as inspirational figures, and named barracks and vessels after them. With the erection of the memorial on the same ground, their importance in history is enhanced further (bmvg 2009: 10).

However, the choice of location sends an additional message. Erecting the monument close to parliament would have visibly expressed the Bundeswehr's role as a parliamentary army, an attribute that German politicians and security scholars have always emphasised when explaining the difference between the Bundeswehr and its antecedents (Mayntz 2008). The placement of the memorial in the MoD courtyard may imply that the Bundeswehr is now a government army, and that the parliament is marginalised (FAZ 2007). This may be seen as an honest reading of a shift in relations between the two arms of state, with the government far more directly involved in running foreign deployments. In fact, the official pamphlet about the monument underlines this factor as a reason for the location (bmvg 2009: 8).

The downside of the location is reduced visibility and use. Among the first to react negatively to the choice were members of the Bundeswehr. Through their intermediary, the Parliamentary Commissioner for the Armed Forces, they let it be known they wanted an easily accessible and centrally located memorial (*Deutsche Welle* 2007). Not surprisingly, some have labelled the end result an 'avoidance monument' (Koldehoff 2009). Apart from the families and friends who go out of their way to visit, all those who pass the memorial do so on their way to or from the Ministry of Defence. This has clear implications for the memorial's use. Foreign delegations visiting the Ministry will pay their respects and place a wreath at the entrance. Since these delegations are composed largely of representatives of the armed forces, the monument has in effect become militarised. When the location debate hit the headlines in 2007, the head of the German Resistance Memorial Centre, Peter Steinbach, warned that if the monument were not given a central location it would end up as a wreath depository (*Deutsche Welle* 2007). That is exactly what has happened. With all the commemoration being done by the armed forces, German and foreign, the monument has become an 'enterprise memorial', with the Bundeswehr as focal point and not the dead (Furhmeister 2009).

Debating the Monument

Making a political survey of the German monument debate is fairly easy when going by the names of participants; their number is limited to the Left faction of the Bundestag, a handful of journalists and even fewer academics. The debate was short lived, probably because of the lack of government response to the criticism. As will be shown below, questions were answered according to protocol, but there was never a comprehensive exchange of opinion allowing the difficult aspects of the proposal to be addressed in any detail. In February 2007, the head of the Bundestag Defence Committee, Ulrike Merten, tried to initiate a debate. She argued that the planning and execution of the monument should be an occasion to make it clear to parliament members and the population that foreign deployments came with a price: 'We have lacked the sufficient explicitness', she added (Güßgen 2007). But no one else was sufficiently exercised to devote time to the monument issue, with the exception of the Left, which already played the main part in debating the war. As one journalist observed, whenever the monument was on the agenda, Bundestag members displayed the same lack of engagement as when Afghanistan was scheduled for debate (Bittner 2007).

Fear of how the electorate would react helps explain such passivity. So does the restrictive planning procedure imposed by the Ministry of Defence. Apart from the announcement by Defence Minister Jung that a memorial would be constructed, public and political involvement was barred. The Ministry invited artists and architects to send proposals for a design, but the jury making the final decision was not named (Güßgen 2007). The selection was presented as if there were no room for changes; the Ministry wished to keep debate to a minimum. Except for the Left and the Greens, nobody in the Bundestag seemed to mind the lack of planning transparency (Weinlein 2007). This was truly exceptional. Berlin is a city where all new memorials have been subjected to protracted and often painful debates. The discussions surrounding the Topography of Terror, an open-air memorial built atop the Gestapo headquarter cellars, is a recent example, but far from the only one (Jaeger 2009).

The Left opposed naming the monument *Ehrenmal*, which directly translated means 'honorary memorial'. Instead they wanted it to be simply a *Denkmal*, or memorial, arguing that there was nothing honourable about dead soldiers (Bundestag 2007). In their request for a name change, they stated that the memorial should embrace other victims than soldiers killed in a war opposed by a majority of Germans. The only appropriate solution would be to have a monument commemorating all war victims, in particular the civilians, who in recent conflicts, including Afghanistan, have died in greater numbers than the soldiers. Development aid workers, policemen and diplomats should also be included (Semler 2007, Nachtwei 2007). That would demilitarise the monument and confirm the post-war civil identity of Germany as a state emphasising development assistance and peace negotiations over the use of military means.

The Left received only a rudimentary reply from the government, stating that all losses were regrettable, and the ministerial plans went ahead (Bundestag 2010a). The party's next strategy was potentially far more damaging. It was to link the memorial with National Socialism. Initially this was done indirectly by claiming that Germany in Afghanistan was serving US imperialism, and that the Bundeswehr ipso facto was an expansionist tool (Jelpke 2009). This was to suggest that those killed in Afghanistan basically were no different from the soldiers of the Third Reich. To highlight the supposed linkage, MPs of the Left asked how many of the dead soldiers had been members of either the Wehrmacht or the Waffen-SS (Bundestag 2009). They knew, as did the authorities, that many soldiers who lost their lives in the early decades after the Bundeswehr's establishment in 1955 had served under Hitler as conscripts or officers (Bundestag 2010b). The government's reply was that it would be unjustified to classify everyone who had belonged to either organisation as a war criminal, and would therefore not pursue the matter any further. The Left's efforts to discredit the Bundeswehr's dead by associating them with Hitler failed to attract much attention, let alone support. The lack of multiparty debate over the monument was indeed remarkable. *Frankfurter Allgemeine Zeitung* concluded that in this respect the parliamentary silence was a sign that 'the post-war era is actually gone, a thing of the past' (FAZ 2004).

The left's arguments were echoed positively by some intellectuals who agreed that the memorial signified a departure from the 'achievements' of modern Germany (Echternkamp 2008, Mittig 2008, Kaiser 2010). The need for a public monument where mourners could gather, a need acknowledged in Britain and Denmark whenever commemorations were discussed, was not given any attention. The focus stayed on ways that such a monument could be combined with modern German political ideals. As *Frankfurter Allgemeine Zeitung* pointed out, it seemed as if the intellectuals were stuck in a post-war frame of mind.

Denmark: Individualising the Loss

In Copenhagen in September 2011, the Queen inaugurated a monument to the country's servicemen and women who have lost their lives in international deployments. The monument had been approved, constructed and unveiled with no German-style controversy. It could nevertheless be seen as a physical expression of Denmark's official framing of the war, much as the Berlin memorial embodies the German government's views and insecurities.

The Copenhagen monument consists of concrete walls that make up three separate enclosures. To the approaching visitor, the inscription 'A time – a place – a man' becomes visible on the outer wall.[3] On the walls of the second enclosure are listed the locations where Danish military personnel have lost their lives. Here, an eternal flame burns. On the walls of smallest, innermost enclosure are inscribed the names and ranks of all those killed, in chronological order. Apart from the open sky, the exterior surroundings are entirely excluded, enabling visitors to concentrate on the columns of names. One striking impression is that the Afghanistan deployment dwarfs all others in terms of the numbers killed. The imbalance of inscribed names and the timing of the monument's construction have made it effectively an Afghanistan tribute. This may change; it is not a final, completed memorial. When discussing the continuity of traditional observances, one usually looks backwards in time to identify a starting point or inspiration. The memorial in Copenhagen is forward-looking. The empty spaces on the walls leave room for new names to be added – a clear sign that Afghanistan will not be the final dangerous deployment for the armed forces, but rather the start of a new era of international engagement.

3 In Danish 'En tid – Et sted – Et menneske'. Although 'menneske' is nowadays usually translated as 'human being', the official homepage of the memorial uses 'man' (Forsvarsministeriet 2011).

When plans for the Danish monument were made public, hardly anybody objected. The few critical comments centred mostly on the site (Faber 2010). The monument stands within the walls of Kastellet, an ancient fortress built to defend Copenhagen from attacking Swedes. It has been a military area since 1624, and although it still is used by the Danish armed forces, the old buildings give it the appearance of an open-air museum. It is a favourite spot for family strolls. The monument is therefore seen and visited by others than those who come specifically to commemorate the dead. The number of visitors has been considerable. Some active-duty soldiers have commented with satisfaction that the memorial resembles the Vietnam Memorial Wall in Washington, a likeness enhanced by the mementoes and flowers left by visitors underneath particular names (Ring 2011).

The monument is not a neutral memorial to the fallen. As in the German case, peace activists and artists interpreted the Copenhagen memorial plan as proof that the armed forces had become an important tool of foreign policy. But ahead of construction there were few protests, apart from the unveiling of a peace monument near the site of a former conscientious objectors' camp near the capital (Christiansen 2008, fred.dk 2008, Avisen.dk 2012). The rival monument failed to attract much attention beyond a few ironic comments in the media. The absence of voices critical to the memorial at Kastellet was remarkable. After all, the recent change in Danish security policy had been implemented with considerable trepidation by politicians who thought that some form of popular backlash was likely.

The absence of significant criticism can be traced to the lack of significant opposition to Danish participation in Afghanistan. Certain aspects of Denmark's role in the Iraq War had caused controversy, but Afghanistan was less contentious, so it was hard to object to the memorial. Potential opponents were also disarmed by the monument's equal embrace of soldiers killed in peacekeeping missions and those who perished in the more recent combat missions. By inscribing all the names, the designers blunted any accusation that they were exalting military adventurism. One might have expected some critics to demand separate forms of commemoration for soldiers killed in UN peacekeeping missions and those who died fighting the Taliban, but no such demand arose. No feature of the monument could be seen as glorification of war or hero worship. There are no echoes of the London Cenotaph's 'Glorious Dead' or even Berlin's 'For Peace, Justice and Freedom'. The monument is to the dead, not their reasons for serving. Nothing, therefore, distracts from the loss inherent in each name.

In its architecture and design, the Danish monument looks beyond the dead as a collective to acknowledge each fallen individual. The third enclosure, with the inscribed names, is the focal point or vortex. The visitor's arrival there is the culmination of a journey that begins with the inscription 'A man' at the entrance. The monument is a physical expression of how the dead have been framed by Denmark's political leaders: as admirable individuals, charged with upholding high ideals. In his speech as the monument was unveiled, the Prime Minister lauded the fallen soldiers for having each made an 'unselfish choice … enabling Denmark to fight for freedom, equality and humanity' (Rasmussen 2011b). By suggesting the war was a question of personal values and national pride, he perpetuated the official framing of the war and its sacrifices. The narrow context left little room for critical voices to debate the on-going war and Danish participation.

Britain: Continuity

The changes we have discussed in regard to Remembrance Sunday have been too subtle to erode the overwhelming impression of continuity through time. Commemorations today centre on the same war memorials that were used before the Afghanistan war. The main difference is that Afghanistan veterans now join the ranks of those who lay wreaths in commemoration of fallen British soldiers. This degree of continuity would have been impossible in Denmark and politically unthinkable in Germany.

The key monument to the dead and the focus of ceremonies on Remembrance Sunday is the Cenotaph in Whitehall. Much has been written about its history and how it came to serve as a model for similar monuments in Britain and around the Empire (Homberger 1976, Greenberg 1989). It is a remarkably simple, unadorned monument. Its elegant form is best appreciated from a distance. Rising from the middle of a busy thoroughfare, this is not a memorial that invites the onlooker to stand still and reflect on any other day than Remembrance Sunday. It has no cross or any other religious symbol. Sir Edward Lutyens, its creator, was a pantheist who objected to promoting any particular religious view at the expense of others. His Cenotaph was a radical departure from traditional war memorials in another way, too. While others sought to justify casualties with inscriptions like 'Dead for the Fatherland', the words on the Cenotaph are simply 'The Glorious Dead'. The inscription does not divert attention from the fallen themselves, yet still manages to endow them with purpose. Their deaths made them glorious. The German memorial does the same, albeit in a more explicit manner, whereas the Danish inscription 'A time – a place – a man' is thoroughly introverted by comparison.

The Cenotaph's inclusivity is the main reason it has come to represent not just the fallen from 1914–1918, but all those killed in World War II and subsequent wars and operations. It does so better than the many monuments whose outward appearance or inscription locks them to a particular event or period (Boorman 2005: 22–3). The Cenotaph, because of its bareness, has remained the focal point for British mourners of all political persuasions. In France, where patriotism and glorification of each era's war was once a prominent feature of monuments to the fallen, a counter-movement has arisen with statues bearing pacifist inscriptions like 'Que maudite soit la guerre' [cursed be the war] (Coëpel 1997). The Cenotaph could not be taken as support for any particular view of World War I, despite what the committee behind it had intended, and this may account for why it soon was copied in other parts of Britain and the Empire.

While the strong element of continuity that lets Britain's military past more or less subsume the dead from Afghanistan may be a source of pride and comfort to recent mourners, it also means that the public perception of fighting in Afghanistan is influenced by previous wars. The practice of adding the names of recent dead to existing monuments further secures Afghanistan's place among military campaigns of history, as does the presence of Afghanistan veterans alongside their elders at commemoration ceremonies. Often the veterans play a highly visibly role in depositing the wreaths (Brooke 2011, Rutherford 2011). In Germany, veterans on the contrary play no part in official ceremonies at all.

Does the addition of Afghanistan casualties to the long honour roll of British war dead deprive the later ones of recognition for their unique contribution and fate? The question has preoccupied some culture historians (Ashplant et al. 2004). Their answer, on the whole, is affirmative. It also seems based more on theoretical reasoning than on the responses of those left behind by the Afghanistan dead. The attitude of family members and soldiers to monuments like the Cenotaph suggests that narrowly focused recognition is not a prime concern. For them, as for their counterparts in Denmark and Germany, the point is to be recognised at all. The parents of a British soldier killed in Afghanistan fought to have his name included on a local war memorial, arguing: 'It doesn't take much at all – just to be included. We're not saying we want to take over or anything like that, just to be acknowledged' (Baker 2010).[4] That appears to square with the basic view of soldiers as expressed in the arrangements for their own funerals that they are asked to make before being deployed. For most, the overriding wish is to preserve their identity as soldiers and members of a team. The anonymity inherent in 'The Glorious Dead' would seem to accomplish that. One must not overlook the importance of soldiers' wishes to the loved ones who are left behind. They seem to find some degree of comfort in the continuity of Britain's recognition practices, including its common embrace of the dead in all wars. Moreover, the inclusive tradition endows the Afghanistan war with the historical gravity the government has been so eager to instil. Ceremonies and monuments are a way of focusing public attention on the war at a time when media coverage has been waning.

Conclusion

Men and women in uniform and their families in all three countries agree on the importance of memorial days and monuments as official forms of recognising of the dead. For politicians, the dead count differently. In Britain and Denmark, deaths in Afghanistan do not contradict the official rationale for deploying force there. In Germany, by contrast, each returning coffin has obscured the official framing of the mission as a development programme. This accounts for the refusal to update the People's Day of Mourning, the removal of the Oath Taking ceremony

4 Eventually, the local council changed its policy and invited the public to send in names of local inhabitants killed in war since 1945 to have them inscribed on the war memorials (Harlow Council 2010).

away from the parliamentary grounds, and the sense of compromise that permeates the memorial erected in front of the Ministry of Defence.

However, there is a caveat here: none of this is permanent. Of course, the German monument is unlikely to be moved to a more prominent location any time soon; nor will Remembrance Sunday cease to be celebrated in Britain. But none of the three countries examined here has settled for all time how to recognise its soldiers killed at war. Their memorial days and monuments have all been created by political fiat, reflecting how politicians wanted to represent the nation and its conflicts at a given point in time (Mosse 1990: 105). The general public's view may be very different. In history, discrepant views have sometimes appeared immediately. The public response to Armistice Day in Britain is a telling example. Initially, two minutes of rejoicing were planned in recognition of the great deliverance of the cease-fire in 1918 (Coppin 1965). This failed; what people wanted was two minutes of shared remembrance, and that is what the event has since become. How the Danish and British publics will mark 5 September and Remembrance Sunday in the future is uncertain. Bitter controversy over the Afghanistan war and its dead could have undermined official commemorations of the future. But so far, there has been none of that in Denmark or Britain. Speaking more generally, the impressive levels of public participation and the widely quoted comments of support from soldiers, veterans and afflicted families would seem to guarantee that commemorations will continue in both countries. Blocking such a clearly expressed public desire to honour the dead would border on arrogance or political folly.

The fact that public attitudes and reactions may differ from what the politicians intend should serve as a warning against exaggerating the political contents of national memorial days and monuments. Jay Winter, who has written one of the few comprehensive analyses of how World War I was commemorated, emphasises the fleeting nature of post-war emotions: 'The dead *were* forgotten; peace did *not* last; memorials faded into the landscape. It is a moot question, at the very least, as to whether healing at the personal level followed' (Winter 2006: 140). I have found little to support this view. Some monuments have indeed been forgotten, but a great many are not, and public participation in the two minutes of silence on Armistice Day in Britain remains impressive. The addition of new names to monuments prevents them from becoming mere historical relics fading 'into the landscape' (BBC 2010).[5] And as far as can be ascertained, none of the monuments were erected with the express goal of preventing new wars, though the emphasis on sorrow and pity may have been intended as a warning to new generations against looking positively upon war and combat. Finally, the wish by families and friends to have the names of their loved ones either on internet memorials or on the plinth of a monument is a warning against discarding the healing effects of public commemorations.

5 Kelly R. Llewellyn (2012) has investigated the motivation for building World War I memorials in Essex and how they have been maintained since. Her findings contradict Winter's conclusions.

Superficially, the main difference between the three countries – or to be more correct between Germany on the one hand and Britain and Denmark on the other – is the way they mourn their dead. In Germany, the People's Day of Mourning is an occasion for lamenting victims without any of the festivity characterising the music and processions staged in Denmark and Britain. All three countries respect the dead; what differs is how they allow the past to impinge on the present. In Germany, the political leadership has taken important steps towards linking the sacrifices in Afghanistan with the political ideals of the Federal Republic. This change was most clearly expressed at certain high-profile funerals. However, it remains politically impossible to incorporate recent deaths into the long-established pattern of official mourning associated with Germany's acknowledged role as the cause of two world wars. Danish and British politicians do not have to confront a similar past. In the Danish case, one may conclude that the celebrations aim at consolidating the country's new security role. Sorrow is mixed with a strong sense of national pride. Amidst the military dirges and sombre poetry, the pride comes through in the politicians' framing of the dead as noble bearers of Danish national values, with a lineage dating back centuries. German leaders by contrast have carefully avoided linkages to history. The Chancellor and Minister of Defence have invariably underlined how the deployment in Afghanistan intimately reflects the values of modern democratic Germany. In Denmark and Britain, veterans hold centre stage in commemorations. An important difference between them is that some of the British veterans date their service to World War II and their numbers are so large that they endow the ceremonies with sense of history impossible in Denmark.

The national differences in commemoration style are reflected in the monuments examined. Nevertheless, behind their variety are several shared features. The most important is the lack of an explicit reference to a given war. That means the focus is directed towards the dead themselves, and less to the circumstances of their passing. The Danish monument comes closest to breaking from this approach, with its list of geographic locations where Danish soldiers have lost their lives. But this appears in the second enclosure, rather than the inner sanctum, and the sheer number of place names prevents a single one from standing out at the expense of the rest. Indeed, without the place names, the monument would have felt even more tied to the Afghan war than it does. In Berlin, the stylised dog tags on the wall of the memorial are a military symbol, but a universal one, not German. Nor is it tied to any war in particular. Another aspect shared by the three national monuments is the timelessness of their frames of reference. The Cenotaph has been transformed into a monument for the British fallen in all wars and military operations since the 1918 ceasefire. The Danish and German monuments for their part are intentionally incomplete, with new names to be added in the future. One difference among the three countries is how the relationship to the state is expressed in commemoration ceremonies. The Cenotaph and the processions on Remembrance Sunday exemplify how close that relationship can be, with the key role of the British monarch underlining the linkage between the armed forces and the nation. The same occurs in Denmark. The contrast with Germany is clear. There, the choice of the Ministry of Defence instead of the Bundestag as the monument's location is an expression of the fact that the armed forces have become an executive organ. It may also serve as a succinct reminder of the Bundestag's lack of involvement in the war and its disinterest in the casualties.

Nevertheless, the German memorial did get built, and as in Denmark, it did not take long. Moreover, the criticism, while loud, was short-lived and limited in scope. Germany's achievement contradicts the conclusion reached by the US historian Michael Keren, who has written extensively on war memorials. He ascribes changes in commemoration practice, war monuments included, to 'decline of the nation-state as the main moulder of virtues' and to 'the objection to war commemoration by feminist groups objecting to the masculine heritage involved a general anti-war sentiment', both of which make it increasingly difficult for veterans to initiate acts of commemoration (Keren 2009: 2). The main problem for veterans in Denmark and Germany has been the lack of a suitable monument on which to focus commemorations, not changes in the nation-state per se. Moreover, the astonishingly short time between the decisions to build the two memorials and their inauguration should be taken as a sign of continued interest by politicians and populations to commemorate wars. German politicians' penchant for unfortunate compromises does not undermine such a conclusion. Veterans in Germany have found it difficult for historical reasons to achieve the recognition they seek, whereas Danish veterans have faced almost no obstacles, for the simple reason that their narrative suits the government's framing of the war in Afghanistan and the soldiers killed there.

Chapter 5
Killed on Our Behalf?

In this book, I have shown how different ways of repatriating and commemorating soldiers killed in Afghanistan have affected European attitudes to the war and the armed services in general. The findings of my study suggest that public support for military goals may be less sensitive to casualties than many political leaders and scholars have come to believe. Furthermore, I question to what extent the war in Afghanistan has affected the public image of soldiers as warriors. In recent years, much has been written about society's supposed drift towards a value set in which soldierly qualities are seen as alien (Cocker 2001, Wevelsiep 2011). The existence of such drift is not supported by my findings, although I agree that there is a gap between men and women in uniform and the rest of society. I will argue that the dead can help narrow that gap, and in doing so give added meaning to their sacrifice.

As outlined in Chapter 1, casualties are often perceived by politicians and security scholars as a threat to any government engaged in military operations. How to frame fatalities in a way that minimises the impact on popular backing is a constant concern. Each government examined here has had its own approach. In Germany, the war in Afghanistan and the German soldiers killed there were such awkward topics to discuss that government and military leaders took refuge in euphemism and kept information about fatalities to a bare minimum. To avoid media attention, the Chancellor stayed away from repatriation ceremonies and funerals until April 2010, eight years after the first coffins had come home. The Danish government chose the opposite approach. Official attendance at every Danish ceremony has been robust, and the media attention correspondingly strong. Denmark's dead were seen as powerful political tools in a redefinition of the country's security policy. In Britain, where the public took some degree of pride in prior military conflicts, the procedures for repatriating, receiving and commemorating dead soldiers were well established. That meant the country's losses in Afghanistan could automatically be framed within a continuing military tradition.

Common to all three countries as the years passed was a growing perception that the war was lost. The political leadership has never openly admitted this. Instead, various attempts have been made to classify the impending withdrawal as the final act of a war that, if it not won, has at least not been in vain. American security experts and political commentators have been the strongest advocates of this interpretation, though there are echoes on the European side of the Atlantic (*Berlingske* 2012, Cordesman 2012, Drury 2012, *Jyllands-Posten* 2012, Nagle 2012). The Danish left-of-centre government elected in 2011, has come the closest to admitting outright defeat, with the Minister of Defence stating in 2013 that the war had never been possible to win militarily (Lund 2013). Like his German and British colleagues, he has pointed to the assistance provided to the Afghan population in terms of education, health care, new roads and a vastly improved electricity system. The Taliban regime has been ousted and elections held. Like them, he has glossed over the fact that the future of all that seems as best precarious once the protective presence of Western forces disappears.

Each country's soldiers have framed what they know about death in Afghanistan in a different way. In all three countries they rapidly became an alternative source of information about what was happening on the ground. What they have said and written in interviews, blog posts and books has undermined and contradicted the official line on important points, forcing their governments to respond. The official line in Copenhagen, however, has never contrasted sharply with the ground view in Afghanistan, and Danish governments have drawn upon the soldiers' accounts to justify its conduct of the war. In Germany, the initial gap in perception was so large that the government eventually was forced to change both the way it described the war and the way it received home its dead soldiers.

More often than anybody else, soldiers have been the ones to pose uncomfortable questions about how political leaders have related to the war and the dead. Their most insistent question has been whether the Afghan cause is worth the lives lost. This is rarely done explicitly; yet when soldiers describe the death of a comrade or recount a farewell ceremony or funeral, they often betray an underlying sense of doubt. Early in the war, when someone died in combat, the soldier was generally described as protecting others in his or her unit – a worthy death. Later, as Western troops began leaving regions that had been gained after protracted fighting, and where the return of the Taliban was considered a strong likelihood, the question of what people were dying for became less easy to answer, not only for the soldiers but for the politicians in charge. Military authorities would often refer to withdrawal operations as strategic adjustments in a planned transfer of control to Afghan forces. Only a few newspapers would see through the spin and call what was happening a withdrawal.

When casualties are discussed, there is a difference in openness between Denmark and Britain on the one side and Germany on the other. British and Danish politicians, military leaders and security experts have openly discussed the utility of force to solve faraway security threats, and the lessons to be drawn from this war in particular (Denning 2010, Ringsmose and Thruelsen 2010, Rasmussen 2011e). Fewer Germans have done the same. Official declarations, such as the 1994 and 2006 white books on German security, state in no uncertain terms that the Bundeswehr may be deployed far from home, but in public debate, the use of military might rarely makes it to the top of the agenda. Nevertheless, the Afghanistan war has made it acceptable for the government to publicly recognise the Bundeswehr as a tool for protecting national interests. When Defence Minister Struck claimed in 2002 that Germany was being defended in the Hindu Kush, it triggered a wave of ridicule and criticism (Struck 2002). In April 2010, Chancellor Merkel repeated Struck's claim three times in a speech to the Bundestag (Merkel 2010). This time there was no derisive response from the political opposition or the media.

Elite perceptions of war, whether shared openly or kept private, are unlikely ever to match what soldiers on the ground experience. The answer to whether casualties are justified depends on who's talking. Service members in Afghanistan have understood their mission to be defeating the Taliban, and to them the job is unfinished (Michelis 2010, *Politiken* 2011). To the governments, the war has always been more than that. British, Danish and German leaders all felt it was important to prove that their countries were credible allies after the attacks on the United States. Both the British and the Danish governments moreover perceived Afghanistan as an opportunity to show that they were international military players. Dead soldiers were an unavoidable consequence.

The question to be asked is whether such activist military policies shine through in the official language used to describe war deaths. In fact, most fallen soldiers are referred to as passive victims of Taliban attacks. And indeed, this characterisation is generally accurate, since roadside bombs and suicide killers have been a major source of deaths. The fact that coalition forces were sent to Afghanistan as combatants is easily overlooked. Only rarely is this aspect considered in the casualty-sensitivity literature. It seems odd, however, to proceed on the basis that the dead soldiers' influence on the debate is separable from what they were doing when struck down. Casualty sensitivity would perhaps be lower if those who lose their lives were not framed as passive victims.

Soldiers as Killers

The German historian Reinhart Koselleck wrote that while 'dying happens alone, killing takes two' (Koselleck and White 2002: 282). Excluding the latter insight obscures what war is about. The degree of official willingness to recognise both the dying and the killing has varied across the three countries. Officials in all three have tended to portray the dead as passive victims. This may be a defensible habit given the prevalence of roadside bombs and suicide bombers blowing themselves up in crowded markets. What's more, accidents outweighed all other causes of death in the first years of the war. Only a minority of fatalities were combat-related. In Germany, such distinctions were never made to the general public, as official communiqués were kept to a strict minimum. In Denmark and Britain, the causes of death have always been listed.

In Britain, the mission of destroying the enemy has never been hidden. The country's long record of warfare, its annual commemoration of fallen soldiers and the widespread media coverage of life under fire in Afghanistan have meant that the dead are framed quite naturally in the context of combat. In Denmark, the combat frame of reference is even more pronounced. Certainly, the success of the film *Armadillo*, with its depiction of Danish soldiers attacking and killing their opponents, has played a key role in making it clear that soldiers also kill. The book *Vi slår ihjel og lever med det* [We Kill and Live With It], did not reach as many Danes as *Armadillo*; but as the work of a serving officer it is a clear example of official openness about military activities in Afghanistan (Møller 2010). Denmark was also the only country where body counts became part of the political agenda, albeit only for a short time.

Germany remains the exception. The dominant official theme of soldiers assisting the Afghan people to reconstruct their society leaves no room for German soldiers in battle. That's why it was a relief to the political elite that the first fatalities occurred in accidents, and not combat. It also explains why upending the official narrative became a key goal of The Left, which tried strenuously to portray dead Bundeswehr soldiers as murderers, even playing the SS card when the memorial in Berlin came under discussion.

The image of fallen soldiers as killers stands in contrast to the image carefully nurtured by the Bundeswehr since its founding in 1955 – the image of citizens in uniform 'winning the peace' [den Frieden gewinnen] as an official slogan would have it. That this Bundeswehr ideal would fall short of what was required in Afghanistan soon became general knowledge. In late 2006, the widely read news magazine *Der Spiegel* (2006) put 'Germans have to learn how to kill' on its cover. The sources for the article were frustrated Bundeswehr soldiers. Only a week later, what would become known as the head-skull affair erupted. The response from the army leadership was to assert the urgency of instilling moral values in the men and women to be deployed (Demmer et al. 2006: 70). The photos of smiling German soldiers posing with skulls became, to borrow the words of US psychologist Stephen Pinker, 'signs of how low our behaviour can sink, and not of how high our standards have risen' (2007: 20). In the resulting political storm, few critics noted that the same army official who had underlined the need for moral values also remarked that the pressure on the soldiers was so great that events like this were no cause for surprise. That would have meant accepting the soldiers' version of what was happening on the ground. Instead, attention remained focused on the damage to the Bundeswehr's image and the country's international standing.

A comparable event was the release of the Danish film *Armadillo*, in which the killing of the opponent was described in triumphant terms that seemed to confirm the worst notions of how war can dehumanise soldiers. Yet, the public response was entirely different from that in Germany. Instead of criticism, expressions of pride were widespread in the media. Part of the explanation is surely that Denmark's deployment had already been framed as warfare, in contrast to Germany's supposed peacekeeping project. Moreover, Danes saw their deployment as an attempt to overcome the pacifism of their recent past, whereas the German deployment was framed by the government as being in line with the country's traditional emphasis on peaceful conflict resolution. For the German government and Bundestag, it was not politically possible to countenance the idea of German soldiers as killers – not after their presence in Afghanistan had been restricted to the peaceful north and they had been barred, as outlined in Chapter 2, from adopting the more belligerent US and British approaches. Germany's reticence was grounded in history. The slogan 'Soldaten sind Mörder' [soldiers are killers] dates from 1931, but has resurfaced at regular intervals in the political debate since World War II. In 1996, the conservative parties CDU and CSU in concert with the Free Democrats proposed a law to 'protect the honour of the Bundeswehr', making it illegal to claim that the soldiers are killers (Hepp and Otto 1996, Spranger 2000). The Left's demonstration in the Bundestag after the Kunduz bombing caused controversy because the gist of the forbidden slogan was repeated, albeit not verbatim. The accusation clashed with the official framing of the Bundeswehr. Whether it clashed with popular perceptions of the Bundeswehr is less certain. The potency of 'Soldiers are killers' has faded as public attitudes have become more realistic. The German debate retains national characteristics separate from those in Britain and Denmark, but the gap is largest when the political levels are compared.

One group that has never tried to gloss over the soldiers' role as killers is the soldiers themselves. On this point, there are no national differences, with the exception of frequent German accounts calling the German mission senseless due to the lack of opportunity to engage the enemy in combat.

Casualties and Public Opinion

Chapter 1 included a survey of the main conclusions of American experts on casualty sensitivity. The extent to which my findings validate these conclusions varies. Studies claiming that there is a direct linkage between shifts in casualty numbers and public support cannot explain the steady drop in popular backing observed in Britain and Germany, and to a lesser extent in Denmark. Moreover, there has been no indication in any of the three countries that milestones – such as 10, 100 or 200 deaths – have triggered a change in popular reaction. One may further question the applicability of the US studies by comparing absolute numbers: in most of the wars cited as empirical evidence, US losses have been infinitely greater than anyone else's. The same goes for Afghanistan, where US casualties by February 2013 amounted to two thirds of total losses (2,177 of 3,257) (iCasualties 2013). Yet, since a key focus of these US studies is the impact of deaths on public support, the authors include interventions with few losses in cases where the dead have made a difference to public backing. Live coverage of dead US soldiers being mutilated and dragged in the streets of Mogadishu in 1993 had a strong influence on public attitudes and political commitment to the operation despite the fact that the losses amounted only to 18 (Dauber 2001). In Britain and Denmark, one may assume that the emotional impact of the Afghanistan war has been heightened considerably by the consistent individualisation of the dead, with their names and faces shown several times and in more than one newspaper. Thus, it may be wrong to assume that simply because the European losses have been small, the emotional impact has been negligible. The question remains as to why my conclusions do not conclusively adhere to the findings of the US literature.

Opinion polls have not played an important part in this book. The general tendency in polling results across the three countries has been to indicate a drop in support for the war. There have been few reversals in the general decline; the most notable was the brief but important hike in British support seen in 2009 (Norton-Taylor et al. 2009). While I have made note of general trends, I have not systematically compared polls. One reason is that pollsters from country to country pose their questions in different ways. In addition, many Europeans appear to have nuanced views on the Afghanistan mission. For example, when those who declare their support for withdrawal are asked follow-up questions – such as whether they would favour continued deployment if a set of specific targets were achieved (better education for girls, improvement in local health facilities, and so on) – the replies make it hard to draw clear conclusions about public opinion. It is no coincidence that John Mueller, the doyen of research on the relationship between casualties and public opinion, emphasises in two recent works the importance of political leadership and communication (1994: 1–11, 2011).

In all three countries the media have played an important part in the debate. Danish and German news organisations have been able to fill a vacuum left by the scarcity of parliamentary debates. In Britain, the media have devoted great attention to soldiers, next-of-kin, military leaders and security experts whose comments are then used by MPs to question the conduct of the war. In much of the literature on the relationship between the media and public debate, it has been assumed that the media influence public opinion. The communication theorist Marshall McLuhan's claim that the Vietnam War was lost in America's living rooms reflects a premise that information flows one way (McCormick 1975), a premise repeated in much of contemporary writing on war and the media (Michalski and Gow 2007, Rid and Hecker 2009). My findings suggest a more complex dynamic. In all three countries studied, the news media have come to rely on men and women in uniform for information and comments about local conditions in Afghanistan. Thanks to social media, those serving in Afghanistan have been able to respond immediately to news coverage at home, and they seem to find it easy to make their voices heard.

The emphasis given to elite cohesion by some casualty sensitivity experts seems to have greater explanatory power when we narrow the definition of 'elite' to signify the political leadership. Included in this group are the main opposition parties of Britain, Denmark and Germany, since they agreed to the war. In none of these countries has leadership cohesion cracked. In Britain, some cracking has been apparent outside of Parliament. Senior officers criticised the lack of adequate equipment, Lord Ashdown condemned the chosen strategy as a failure, and a parliamentary sub-committee concluded that the government had failed to inform Parliament and the public about changes to the British mission. How these cracks have influenced popular support is impossible to say. No opinion pollsters have asked people whether their support for the war was influenced by the officers' discontent, Lord Ashdown's criticism or the sub-committee's conclusions. Some impact is more than likely, but the decline in support that set in soon after the deployment began, seems not to have accelerated due to any of these factors.

The British debate suggests that we might do better to shift analytic focus away from the degree of elite cohesion to the degree of elite commitment. Are the opinion makers passively acceptant, or explicitly supportive? Intensity of support for government policy is difficult to measure, since any military leader dissatisfied with his equipment may lash out at the government for failing to allocate the means to victory. The numerous examples of this in Britain may have coloured the elite response to casualties, though the effect is hard to gauge. All three countries have heard complaints at times about inadequate equipment, rigid rules of engagement and the failure of political leaders to express support for the troops. Yet it is in their response to such accusations that the level of commitment to the war is vividly expressed. Danish Minister of Defence Søren Gade's vocal support for the troops and high media profile meant that, despite the complaints, nobody doubted the Danish government's determination. Only towards the end of the war, when an exit date had been set and few were prophesying military victory, could dissenting voices be heard in Britain and Denmark. The British ambassador to Kabul stated that Western commitment to Afghanistan had been insufficient (Rayment 2012). In Denmark, the centre-left government that took office in 2011 stated bluntly that the war had always been militarily unwinnable (Lund 2013). This was rebutted by a vast majority of those on the ground in Afghanistan, who told the press they would like to stay on and see the 'job finished' (*Politiken* 2011). This discord was an exception in a long record of harmony between service members and the government in Copenhagen. Neither the Danish tiff nor the British ambassador's statement triggered much public response. Had they come earlier in the war, before the withdrawal date was set, more of a storm would likely have resulted.

At times, the political leadership may be more hesitant than the population when confronted with a threat scenario, simply because it has access to more information and therefore a fairer chance of estimating the possible costs involved. Thus, according to the two US security scholars Michael Alvarez and John Brehm, casualty shyness by governments should not be understood only as a knee-jerk fear of public backlash (2002: 194–215). Without access to governmental meeting records, it is impossible to say whether this observation is valid in the three countries studied here. The German government's efforts to keep the Bundeswehr out of dangerous areas and situations in Afghanistan may be used in support of the US scholars' conclusion, although the German policies may also have been motivated by a wish not to be too closely associated with the US, and in particular with President Bush's war on terror. However, Denmark and Britain are examples of the opposite – of perseverance despite heavy losses and no clear victory in sight. The fact that no anti-war movement strong enough to trigger a change in official behaviour emerged in either country suggests that casualty sensitivity alone will not trigger opposition. In other words, numbers alone explain very little. In fact, the conventional wisdom that the population will not tolerate casualties, so widespread in US literature on the topic, seems unfounded. Peter D. Feaver and Christopher Gelpi, who have published extensively on US civilian–military relations, have pointed to the possibility that the public may be far more accepting than conventional wisdom has concluded (2004: 102). That would discredit not only much of the US literature, but also the widespread European belief that the continent's preference for peaceful conflict resolution is a reaction to its bloody past. Denmark and Britain can be used as arguments against a supposed European aversion to war. Even the assertion that populations will be particularly apprehensive when victory prospects are dismal – that most people are 'defeat phobic' – seems not to apply (Eichenberg 2005). In Britain and Denmark, support has dropped slowly, with the already-mentioned exception of a 2009 blip in British support following a sudden surge in casualty numbers. No sharp fall in support has been recorded as the withdrawal date has moved closer. The public's relative constancy is not likely due to the government's official optimism about Afghanistan's future – there have been too many soldiers criticising the withdrawal plans, too many Afghan soldiers attacking their Western colleagues and too many Taliban attacks inside Kabul and other areas formally under allied control. With the withdrawal date firmly set, some people may have found it easier to repeat the official line and answer 'Yes' to the question of whether they favour a pull-out. In fact, how people react to the looming defeat depends partly on what the war was nominally about in their country – in other words, how it was framed for them. It's no surprise that in Denmark, where the war was not only about defeating the Taliban, but also about regaining national honour, support seems to have dropped less. When the Danish government in 2013 acknowledged that the war had not been won, there was no discernible impact on the media debate or public attitudes (Lund 2013). The reaction might well have been very different if this admission had been made at an earlier stage in the war. One important conclusion that can be drawn from Denmark's example is that the notion of success is a flexible one. The same observation had been made about the US public's perceptions of the Iraq War (Gelpi et al. 2009: 188–235). It all depends on how the war is framed.

In Denmark, the debate has been less about what was really gained or lost in Afghanistan than about Denmark's own change in foreign policy orientation. But only a few critics have tried to exploit this insight as a way to claim that Danish soldiers have been sacrificed to gain international influence, and the response has been dismal (Dalager 2010, Jensen 2013, Larsen 2013).[1] Indeed, it would be an exaggeration to say they have stirred debate.

Much casualty-sensitivity literature is premised on the ability of people to balance costs and benefits – that is, human casualties against improved security or other national interests. Few have disputed the validity of this premise, although

1 A notable exception was the article by an Afghanistan veteran asking why religious extremism was left alone in Denmark, when exterminating it had been a prime motive in Afghanistan (Nielsen 2012).

political preferences may enter into the assessment along with information that is framed in such a way that disturbing facts are left out. The reason may well be that we have no alternative. People have to make up their minds on the basis of what they see, read and hear, and that does not immediately translate into one view or the other. Both in Britain and Denmark, opinions polls have shown what appears to be a contradiction: that a majority of people believe the price paid in Afghanistan exceeds what has been gained, while at the same time they support continued deployment (Glover 2009, Kristeligt Dagblad 2011). It may also be the case that Europeans are less casualty sensitive than Americans simply because the European way of warfare has been less high-tech. Only the US's unique military might has made it possible to avoid casualties.[2] Another explanation is that the alternative of not responding militarily to the Taliban did not have much support either. In none of the countries studied was there much debate about the basic approach. Even in Britain, where questions were raised over the conduct of the war and the possibility that better equipment could have saved lives, the need to defeat the Taliban militarily was not questioned. The few members of the Bundestag who would have preferred a more proactive Bundeswehr role did not mount a serious threat to the chosen approach. The balancing of costs and benefits, sometimes offered as an explanation when public backing disappears, was never conducted in the open. That is not to say people did not make their calculations in private, but the lack of any opposition to the official line certainly made it hard to mobilise against participating in the war (Newman and Perloff 2004: 30).

The belief that the public is casualty-phobic, unfounded or not, may become a self-fulfilling prophesy (Kull and Destler 1999, Kull et al. 2003–2004, Feaver and Gelpi 2004: 104). Over-worried about public support, the political and military leadership may choose an approach that is more cautious than public attitudes would actually prefer (Hyde 2000). The implications should not be underestimated: the public may come to believe that future wars can be fought without bloodshed and that participation in a military operation can be conducted without combat. This is not an unavoidable trap. A committed leadership can challenge public sensitivities. This is what Danish Minister of Defence Søren Gade did early in the Afghanistan war when he stated that Denmark was in for the long haul and that more casualties could be expected (Gade and Tørnæs 2006). The German Chancellor and Minister of Defence made an important step towards confronting casualty sensitivity when they admitted that Germany's lost soldiers had been killed in a war. By Danish standards, this acknowledgement may seem slight, but in light of German political culture it was substantial indeed.

---

2 US reluctance to deploy ground troops to drive out the Serbs from Kosovo in 2001 is often cited as an example of the different US and European approaches (for instance Smith et al. 2001). What should not be overlooked is the US perception that this was a war with no bearing on US interests and the willingness to risk US lives was correspondingly low.

Despite national differences, the governments in the three countries have maintained that sending soldiers to Afghanistan was a matter of necessity. Had they said it was a matter of choice, public response to the casualties might well have been very different (Haass 2009). Debate over whether this was a war of choice or necessity never emerged in Britain and Denmark. It could be that the status of the Afghanistan mission was enhanced by comparison with the war in Iraq, where, some critics said, participation was largely a matter of trying to curry favour with the United States.

Dead Soldiers in a Post-Heroic Society

One argument often proffered to explain increased casualty sensitivity is the general depreciation of values traditionally associated with soldiers. Courage, physical strength and determination – warrior values – have fallen into disrepute and been overtaken by more 'feminine' values such as empathy and willingness to compromise. A general shift in Western society may be the cause (Smith and Tatalovich 2003, Farrell 2005: 173–87, Wevelsiep 2011). So long at peace, the reasoning goes, Western societies no longer can muster the requisite number of men with combat qualities to ensure prolonged fighting. Pacifism has gained sufficient hold of the mindset to preclude anything action but peacekeeping. Technological advances have led to the unrealistic notion that bloodshed is avoidable (Wall 2012). Instead of fighting, staying 'at home with the kids' seems a preferable option (Luttwak 1994, Coker 2001: 56–66, 83–8, Naumann 2005, Hettling 2008).

A more sophisticated and less provocative explanation has been offered by both John Mueller (2004) and Stephen Pinker (2011), who conclude that political development has marginalised war. As long as they refer only to major wars, their conclusion is borne out by statistics (SIPRI 2012: Armed conflict). However, the number of armed conflicts has remained remarkably stable since the beginning of the millennium.[3] In recent years, NATO's bombardment of government forces in Libya in 2011 and the French-led campaign against Islamic militants in Mali in 2013 are pertinent reminders that military means will be used in the future to solve conflicts. What these two operations share with most other Western-led operations is an extremely low number of casualties. In Libya, the only loss of British life happened in an accident. The difference between most such operations and true wars is the number of casualties. The war in Afghanistan dwarfs all these military interventions and may therefore well serve as a turning point in the way Western populations perceive their armed forces, including an understanding that deploying men and women to a combat zone will entail losses. This is a particularly pertinent reminder at a time when technological advances in drone aircraft and other fields may give the false impression that future wars can be conducted with

3 SIPRI 2012: Figure 2.2 Number of armed conflicts, non-state conflicts and one-sided violence, 2011–2010.

minimal losses to the West. It is easily forgotten that 'Western' in this context often means American. It has been pointed out that if European forces had not had the technological support of their US ally in the Libyan campaign, ground troops would have had to be deployed, with casualties as an unavoidable result (Schake 2012). It cannot be determined whether the European countries that participated in NATO's bombardment of Libya would have agreed to the use of ground forces. However, none of my findings reveal a profound aversion either to the use of military means or to casualties, even if such an aversion had seemed to have been an integrated part of national culture in the cases of Denmark and Germany.

Much has been written about how values have changed, undermining support for military operations and reducing the public's willingness to let its sons and daughters join the armed forces. But the Afghanistan war may suggest a firming of traditional values. In Denmark, recruitment numbers reached an all-time high during the war (Damm 2010). In Germany, too, the number of recruits completing their service seemed unperturbed by the ongoing war (*Deutschlandfunk* 2011, *Focus* 2011). Thus, Denmark at least may serve as a repudiation of any conclusion that contemporary society has no room for the traditional soldier. So does Germany, if less explicitly. Once the official framing of the German deployment as one of help and assistance became untenable and the political elite had to face up to realities on the ground, remarkably few Germans protested. It should be recalled that this was a profoundly unpopular war in Germany. But apart from objections by The Left and some political scientists and intellectuals who correctly saw the deployment as a step towards greater acceptance of military force in foreign policy, there was at no point a strong, coherent demand for German withdrawal.

What did happen, when Germany's political leadership finally acknowledged the soldiers' version of events, was that the official language changed. The framing of the mission became more like that of Britain and Denmark. It was not just a matter of openly admitting that German forces were involved in a war, significant as that was, but of the whole range of terms used to describe the dead and the living.

When soldiers of any nationality characterise a dead comrade, their words often seem strangely out-dated. Heroism, sacrifice, patriotism: such terms recall an era whose emotions, acts and attitudes are somehow ill-suited to modern society. But the similarity of rhetoric across the three countries, even to the point of monotony, suggests a certain validity in the time-honoured word choices. The soldiers were heroic. They were patriots defending national values, and their deaths should be understood as sacrifice.

In Germany, the dead soldiers are not idealised. The reason is, according to the parliamentary ombudsman for the armed forces, a 'fear of history' (Jungholt 2009). No echoes of Third Reich hero worship can be permitted. Yet some have wondered whether a modern democracy like Germany may not need heroes after all (Davis 2009). Although nobody within the Bundeswehr have advocated this, at least not openly, greater political support for the fighting Bundeswehr soldier would diminish the gap between those deployed and the people and politicians back home.

As far as can be ascertained from written sources, Danish soldiers feel that the job they have done has been sufficiently recognised by official support and popular backing. The men and women serving have become a symbol of national pride – a remarkably quick change in status that dates from 1993, when one Danish naval vessel set off timidly to participate in the first Gulf War. Some of today's soldiers can recall when wearing a uniform in public would attract negative comments (Birkeland 2010: 37). The Danish Defence Minister Søren Gade expressed it thus:

> [T]he conditions in Afghanistan have an effect on us as fellow human beings. As human beings we have to intervene to stop injustice and unfairness. This is the way most of us have been raised at home. It applies to each of us as individuals, and it applies to Denmark as a nation. It is required that someone makes the sacrifice and pick up the burden. (Gade 2009)

In this conception, a military or strategic debate is reduced to one of morals. The implication is that only chaps with moral fibre can do the task – and needless to say, they represent something admirable. Moreover, by stating that the burden is there to be taken up, Gade emphasises individual responsibility over any mandate thrust onto the soldiers by politicians. Interpreted cynically, it is a transfer of responsibility from the government to the uniformed services. But making a war into a personal choice about morals also makes it more difficult to criticise or oppose. This rhetorical approach may help explain why public support for the Afghanistan war remained higher and dropped far more slowly in Denmark than in Britain or Germany. British politicians used similar moral language from time to time, especially in describing the sacrifice of fallen soldiers. But they never adopted the fervour of their counterparts in Denmark, who knew that support for Danish troops was a key to consolidating the country's new security role. Nor were the British as overt as the Danes in portraying soldiers as the bearers of national values. Deployment abroad was after all a well-established British practice and less controversial than in Denmark. Only in Denmark would a foreign minister equate criticism of the war with support for the Taliban (*Jyllands-Posten* 2009).

Making the war into a moral issue, implying that those against the war lacked the moral fibre of those serving, was never attempted in Germany. Indeed, the statement from General Kneip quoted on page 81 showed an automatic rejection of laudatory terms (Meyer 2011). The old slogan that soldiers were 'citizens in uniform' was taken to mean that nothing, not even courage and moral fibre, distinguishes people serving in the military. Even when Germany's medal for courage was handed out, the event was downplayed in the media and the ceremony conducted behind closed doors. The paradox is that by underplaying the gap between soldiers and citizens the authorities have actually widened it, encouraging a distaste for military values that feels to some soldiers like a form of ostracism.

For them, the gap between life at the front and at home is often reduced to the fact that no one at home understands what it means to be confronted with death continually. Soldiers often refer to death as 'the highest price' or 'the ultimate sacrifice'. In both Denmark and Germany, commentators have wondered whether such venerable terms remain valid today. But the fact that soldiers are comfortable speaking of the heroism and sacrifice of their comrades is proof the words have not lost all salience. In Britain, politicians and the press have used such language from the start when referring to the dead of Afghanistan. In Denmark and Germany, the public vocabulary has changed gradually, catalysed by the burials of soldiers. An examination of newspapers reports and TV coverage of these events shows a marked change in rhetoric since the first bodies were sent home.

Those who conclude that Western European countries have become casualty sensitive to the extent that warfare is impossible often turn to European history for an explanation. The fact that continental Europe has been ravaged by two world wars is often cited. The prime example is Germany, where the main legacy of the Third Reich is a political culture marked by pacifism. In debates over the Afghan war and how to receive home the dead, Germany, the UK and Denmark view the past very differently.

In the British debate, explicit historical references are rarely invoked, not even the British Empire's two failed attempts to subjugate Afghanistan in the nineteenth century. When the current war began, Prime Minister Blair's only nods to history were general, bordering on vague. They concerned the readiness of the armed forces to defend British interests. Denmark's leaders took a different tack, at times explicitly referring to history to justify participation. In Germany, almost every time an official statement or declaration was made about the deployment to Afghanistan, the past would be recalled, although it was never the same one. The government would point to the post-war period, with its emphasis on democratic values and alliance membership. Opponents of the war would reach further back and compare the Afghanistan mission to the militarist aggressions of the Third Reich. Even German academic literature referred in great detail to the long record of foreign defeats in Afghanistan, using it as an argument against attempting anything as foolhardy again. As pointed out in Chapter 2, Germany had better relations with Afghanistan and more in-depth knowledge about it than either Britain or Denmark did when the war started.

If we accept that the past is the main explanation for the differences between the three countries – with pacifistic Germany on one side and Britain at the other extreme with relatively few inhibitions against using the armed forces as a political tool – then it would appear safe to say the past is quite likely the key to the behaviours seen in Afghanistan. One should nonetheless be wary of regarding the past as a political straitjacket. Denmark may serve as the clearest evidence that this is not the case, but the significant changes initiated in Germany may also serve as a reminder. The difference between these two countries, one with great explanatory power, is that Denmark's change of course was politically initiated before the war started while Germany's came in midstream, when the political leadership finally realised it was out of touch.

Political and Popular Commemoration

As shown in Chapter 4, commemoration of the dead is important to the next-of-kin and fellow soldiers as an official sign of acknowledgment; for society at large, commemoration makes it possible to express public support for grieving families and for the men and women still in uniform. What should not be overlooked is that parades, flag days and monuments enable populations at home to come to terms with a distant war.

Yet as this book has shown, not all dead soldiers have been publicly grieved or commemorated. It is true that soldiers from all three countries give similar accounts about death in Afghanistan, and there have been few national differences in the base ceremonies conducted before bodies are sent home. But differences do emerge when the dead arrive in their home countries. The contrast between the German policy of silence, which lasted until 2010, and the British and Danish openness could not have been stronger. All evidence suggests that soldier deaths are 'grievable' (to paraphrase Judith Butler) only when the general public is given the opportunity to mourn, and for that to happen political commitment to the soldiers is a requisite (Butler 2010). In a survey article on casualty aversion, Hugh Smith concluded: 'Accepting private grief for the public good is less common and more difficult than in the past' (Smith 2005: 503). Notwithstanding the difficulty of documenting whether people in the past were more accepting of private loss, the sources presented in this book indicate that public commemorations are a great help to those left behind. To the armed forces, public attendance is taken as a sign of popular backing.

The strong attendance at commemoration ceremonies in Britain and Denmark indicates that the populations there value such events as a way to express support for men and women in military service. To equate attendance with support for the war in Afghanistan would be wrong. In fact, little is known about why civilians crowd the pavements when coffins are transported through town or a homecoming parade takes place apart from a wish to pay their respect to the dead. Many would probably recoil from explicitly political messages that might engender controversy. This is in line with conclusions drawn by the British sociologist Paul Connerton in his book *How Societies Remember* (1989: 45). Effective commemorations allow everyone to decode the day's ritual into a private language, Connerton concludes. If wartime rituals are overtly politicised, the scope for interpretation narrows and those who would like to express respect for the dead while at the same time disagreeing with the war would likely stay away (Connerton 1989: 53). When the British Legion in Royal Wootton Bassett stated they did not want politicians making speeches during the repatriation ceremonies, the reason given was that it would be disturbing (British Legion 2013). Also in Denmark, the political content of commemoration ceremonies was made less explicit through the choice of date. Although the date chosen marks a watershed in Danish security policy, its significance is hardly known outside the small circle of security experts.

The verbal restraint by government members or heads of state should not detract attention from the fact that ritual itself, its form, conveys an unmistakable political message. That message is unique to each country. In Britain, the emphasis on tradition has allowed the Afghanistan dead to be welcomed into the larger community of fallen heroes dating back to World War I. The remembrance ceremonies in Britain are remarkable for their permanence of form. Despite entertainment shows on television the evening before Remembrance Sunday, the ceremonies in Whitehall have changed very little. The strong degree of continuity, the re-enactment of cherished procedures, is part of what gives the ceremonies their meaning. It also shapes collective memory (Halbwachs 1997).

In the two other countries examined here, there was no tradition to build on as the bodies began returning from Afghanistan. In Germany, a semblance of continuity might have been achieved by repurposing the People's Day of Commemoration. Doing so was never a topic for discussion; the day remained the nationwide occasion for remembering not only the losses of the two world wars, but also Germany's role as perpetrator of war crimes. Updating the People's Day of Commemoration to include Afghanistan would have been seen as an attempt to diminish the event's penitential function; it would also make it too easy for critics to equate today's soldiers with Hitler's Wehrmacht. So for now Germany does not have a public occasion for mourning its recent losses. Until 2010, when the Chancellor and the Minister of Defence acknowledged the war-like nature of the mission in Afghanistan, it was perhaps tenable to argue that no such occasion was called for. Since 2010, the reason for doing nothing is political trepidation over how the public will respond. As a result, the public has no opportunity to pay its respects. This means that the families of the dead and the Bundeswehr as an organisation have to commemorate their losses in private, either within family circles or the confines of the military bases where commemoration services have always been held. The monument in Berlin that's open to the public seems destined to fail in its role as a centre for public commemoration, both because of its locality and the partial anonymisation of the victims. The importance attributed to having the names imprinted permanently, for all to see, comes through in most personal accounts, interviews and articles addressing the issue of commemoration. When comparing the German monument to those in Britain and Denmark, what stands out as unique is the German focus on the political ideals of the Bundeswehr, rather than the lives lost.

A comparison of the three monuments elicits another, rarely expressed aspect of successful commemoration whose importance was not clear to me before visiting the site in Berlin: public visibility. Unless people can get to a monument easily, as with the London Cenotaph, its effect on collective remembrance will be negligible. The French historian Pierre Nora, who has written extensively on how the past is remembered, has concluded that modern memory 'is above all archival. It relies entirely on the materiality of the trace, the immediacy of the recording, the visibility of the image' (Nora 1989: 13). A monument that cannot be seen, like a commemoration ceremony restricted to the few, does little to preserve public memory.

It should be added, though, that no monument guarantees profound reflection. The American artist Edward Kienholtz illustrated this problem ironically with his Portable War Memorial, a meticulous copy of the famous Iwo Jima statue, with dotted lines where anyone could write in the war of their choosing (Kienholtz 1996). His point was that a memorial by itself is empty, and can just as well be used for lauding militarism as anything else. Yet a monument that is out of sight seems destined to play no role whatsoever. The Berlin monument may serve as a short-lived exception. It did trigger several critical questions about the Afghan war, the treatment of dead soldiers and modern German political culture when its location and contents were on the agenda in 2009. Since then, the monument has barely been mentioned.

It may seem paradoxical that while support for the war has dropped in both Britain and Denmark, public attendance at commemoration ceremonies has grown. Public displays of support and respect are numerous. In Denmark, part of the explanation may be that the government has succeeded in framing its soldiers as admirable Danes restoring the country's military honour. In Germany, local funerals are the only opportunity the public has to show respect for the dead. Whether public ceremonies in that country would generate Danish-level attendance cannot be said. The plan for a German Veterans Day similar to British Armed Forces Day has been stalled, and the ceremonial oath-taking by German recruits is no longer a public event, so comparisons with Denmark and Britain are impossible. The only indication that the broad German public has high regard for the armed forces comes out in annual surveys. The Bundeswehr ranks consistently high – and always higher than the politicians (Keil 2010).

Without a memorial day to commemorate the Afghan dead, an opportunity to debate and reflect on the meaning of German casualties is missed. When the Bundeswehr Centre for Leadership Development and Civic Education issued a working paper in August 2010 on the need for more official openness when the Afghanistan war was discussed, the need to talk about death without any taboos was underlined repeatedly.[4] The paper was not made public. It does seem to me that this debate is bound to find a venue, even if the politicians are reluctant to address the issue. Servicemen and servicewomen coming home will continue to make their views heard. The same is true in the two other countries; but while in Britain and Denmark the returnees seem focused on communicating with the general public about what Afghanistan was like, the target audience in Germany seems more to be the political leadership.

Soldiers' recollections of Afghanistan will change in the years ahead, much as the memories of past wars have evolved. It may be expected that many will question the official justification for the war and ask what was gained. In Germany, such questioning has already occurred, but in Denmark and Britain few of the books published have had a critical perspective.[5]

4 The paper is not publicly accessible. For an extensive summary see Birta 2010.

5 Among the exceptions are Rasmussen 2011 and Townsend 2012.

In all three countries, the official line has been that the war has been a success insofar as it has prevented Afghanistan from being used as a safe haven for terrorists. If the vacuum left by Western soldiers is filled by the Taliban, this justification will ring hollow (*Politiken* 2011, Nicol 2013). It may also force the question of whether the soldiers have died in vain. How governments in retrospect justify the lives lost cannot be isolated from the official treatment given to fallen soldiers over the years. The Danish government may be better prepared than the others to face the question of what was achieved, since it has now stated that the war was unwinnable and that Afghanistan remains a fragile state (*Information* 2012, Hvass 2013, Lund 2013). In the official *Afghanistanplan*, issued by the Danish foreign and defence ministries in 2013 to outline the reduction in Denmark's military presence, the possibility of setbacks and failures is mentioned on almost every page (Udenrigsministeriet 2013). Nothing comparable has been issued in Britain or in Germany.

Apart from the emotional aspects of commemoration, the monuments and ceremonies of the three nations have a political role that is not always immediately perceptible. As focal points of public attention, they may give rise to reflection over what the wars being commemorated mean to national identity. The German People's Day of Mourning is a good example of how this has been done in the past; another is the Danish and British prime ministers' speeches at annual commemoration ceremonies. No matter how repetitive the events may seem, they are an official acknowledgment of responsibility for the dead soldiers. They are also a meeting point between the armed forces and the rest of society. As such, they can help close the gap that has widened between the two groups since the end of the Cold War.

Closing the Gap

This gap has been a topic of much research on both sides of the Atlantic. US and European researchers conclude that, unless efforts are taken to bridge it, alienation will grow to the extent that communication will be impeded and national security will suffer (Feaver et al. 2001, Wevelsiep 2011).

The German discussion about the civilian–military gap has had a different starting point from the same issue in British media and security literature (Denning 2010). In Germany, the assumption that the Bundeswehr is an integrated part of society makes the slogan 'civilian in uniform' highly relevant. The concept of everyman-soldiers and the fact that their sole mission, as originally conceived, was home defence underpinned a so-called *Gesellschaftsvertrag* [societal agreement]. Some German security scholars have concluded that military deployments have since broken this agreement (Bake 2010, Meyer 2010, Müller et al. 2010, Mannitz 2011). That is true only insofar as both parties to the pact (the parliament and the armed forces) agreed that it could not be altered. This does not seem to have been the case. Like other Western forces, the Bundeswehr changed its training regimen after the end of the Cold War to accommodate the increased likelihood of foreign deployment. The Ministry of Defence stated this in no uncertain terms in security white books published in 1994 and 2006. However, Germany debated the shift less openly than its allies did. In a large research project undertaken by the Peace Research Institute Frankfurt from 2006 to 2009 – during an early phase in the Afghan engagement – German soldiers revealed that talking about the possibility of dying in combat or killing the enemy remained taboo, even though that was what defined a soldier's role (Bake 2010: 132–3, Müller et al. 2010).

According to the historian Paul Nolte, Germany's soldiers are perceived as 'alien' to the country's political culture (Nolte 2008: 184). Deployment abroad is therefore ideal because it means that the armed forces are used, but without catching the public eye (Kaim 2009). The armed forces have in Nolte's view become invisible. With national conscription disbanded, fewer and fewer Germans have any personal experience of the armed forces. Trouble, of course, arises when the deployed men and women return in coffins. Then 'invisibility' is no longer a tenable strategy. It should not be overlooked that once the political behaviour of Germany's leaders changed, and the dead were described as heroes with all of Germany bowing its head, there were hardly any objections (FAZ 2010). If the population truly perceives the armed forces as alien, as Nolte and many with him have concluded, that would not have been the case.

When Angela Merkel praises German soldiers for their bravery and states that they are models for the rest of society, she breaks a taboo by implying that people in uniform are different from everyone else. Apparently they are no longer just 'civilians in uniform'. When her British and Danish colleagues suggest the same thing, no one bats an eye. Merkel's belated tributes contrast with the timid statements issued previously by German authorities as well as the many years of refusing to acknowledge Afghanistan as a war zone and the Chancellor's own absence at early funerals. For Germany, the change has been momentous.

The gap between the armed forces and the rest of society is less fundamental in Britain and Denmark. There, the use of the military is not perceived as alien, but is rather an accepted part of the states' instruments. The gap is experienced more as a challenge in communicating what the war in Afghanistan been like for British and Danish troops. When a large exhibition on the Danish military experience in Afghanistan opened in Copenhagen in late 2011 under the title *Den fjerne krig* [The distant war], even those in favour of Danish participation pointed out that despite the realism conveyed by the sandbags and ramshackle latrines, the exhibition had little to say about the human costs involved (Knudsen 2011, Olesen 2011). Trying to capture what the war was like, and communicate it to a wider public, is a challenge confronting all returning soldiers. The Copenhagen exhibition and the British Imperial War Museum's exhibition of reflections by Afghanistan war veterans, held in late 2011, were both attempts to close the gap between service members and civilian society (Imperial War Museum 2011–2012). No German museum has staged anything similar. As far as can be ascertained, the only German exhibition about Afghanistan was done by a private, charitable organisation and focused on the war's impact on the civilian population (Rusch 2010).

The Political Value of Dead Soldiers

Dead soldiers have been regarded in all the casualty literature as a burden to political leaders. The lesson from Vietnam was that body bags undermined popular backing to the extent that it threatened the government's ability to conduct war. If later forgotten, the lesson was relearned during the botched Somalia operation in the early 1990s. Afghanistan suggests that it is not necessarily valid today. Both in Denmark and Germany, caskets returning home went from being a political burden to a sign that the past could be left behind. For the dead to have positive political value, the war's framing must be credible, and its rationale widely accepted. If the framing is off, as when German political leaders tried to convince the electorate and the media that the deployment was about something other than conflict, the dead can be only a burden. They become the ultimate proof that the government's strategy has failed.

The political value of dead soldiers, positive or negative, is above all influenced by the war in which they die. Afghanistan has ended in defeat. None of the political targets defined at the onset have been achieved. Even the very significant progress made in health and education will be difficult to sustain without the presence of Western troops. Thus, the political value of the soldiers' sacrifices is coloured not only by the war's framing, but by its result. The impact of the withdrawal is felt differently across the three countries. It depends on their initial targets. For Germany, few of its stated goals have been achieved. It is nearly impossible to find out about the assistance and development projects that loomed so large in those early government reports to the Bundestag. Had they been a success, information would likely be available. In a parliamentary brief issued by the governing CDU faction in parliament on the reduction in Germany's presence in Afghanistan, civilian assistance is barely mentioned (Schockenhoff 2012). Denmark finds itself in a different situation, because it had different goals. Defeating the Taliban was just one of them. Another was proving that Denmark is a military power willing to fight. Seen from this perspective, Denmark's dead are the by-product of a new and hard-won international prominence. In Britain the debate over fatalities has so far remained focused of matters of equipment and to a lesser degree strategy. The controversy over whether more helicopters would have saved might easily have turned Britain's dead into a heavy political burden for the government. But since the war has been seen both as a confirmation of Britain's role as an international political actor and of the continued importance of the close relationship with the US, the deaths are accepted as unavoidable. The fact that military losses are seen as part of a long and glorious British tradition has also lessened the impact.

The national filter, so distinct when each country's political leadership refers to its dead, is far less noticeable when withdrawal is being addressed. The decision to pull out of Afghanistan is usually portrayed as coming from NATO or the United States (Cameron 2011). The question of who bears responsibility for the outcome of the war is thus blurred. Denmark provides an example of how a government can pass blame for the loss of life to NATO. In 2009, an officer claimed publicly that soldiers had died in vain because of the government's handling of the war (Lahrmann 2009). The media found a few other Danish officers who supported him, including one who added that attending funerals was difficult without having faith in the mission's success (*Politiken* 2009). The government's response was that expressing doubt only benefitted the Taliban (*Jyllands-Posten* 2009). This was potentially a dangerous debate. The admixture of an official policy in transition, lack of progress in the mission and dissatisfaction on the ground could have sparked an overall debate on what could realistically be achieved. No such debate occurred, mainly because the political opposition chose to hold its fire but also because of the correction from the officers' organisation. It was asserted that the problems were due to the lack of a coherent strategy for the Western mission. Responsibility was not at the national level, but higher up.

That being said, the sense of defeat has certainly not been equally palatable in all three countries. In Germany, the defeat may well lead to a reinforcement of the old pacifist approach to security politics. A military solution has failed to bring peace and stability to Afghanistan, and insurgent activity that had mainly been confined to the southern regions and the capital has spread with the increase in US, British and Danish attacks on the insurgents. The German approach may seem vindicated. It is not difficult to find recommendations from prominent scholars, politicians and journalists for a return to the traditional German approach to security politics (Rinke 2010, Rode 2010, Triebel 2010, Graw 2012, Hippler 2012). The refusal to participate in Libya and the belated offer of logistical assistance to France during the Mali intervention, days after Denmark and Britain made concrete offers, seem to indicate that Germany has not got over its aversion to military activity (Seibel 2013). What Afghanistan has shown, on the other hand, is that Germans in general may be less nervous about combat and fatalities than the political elite in Berlin. The gap between returning soldiers and society at large may be smaller than the gap between the soldiers and the elite.

The political value of war fatalities depends very much on how the war in question is perceived by the public. At issue is whether the cost, in bodies, outweighs what has been gained (Betts 1999: 167). The answer varies by nationality. In Denmark, the fallen soldiers of Afghanistan loom much larger in the political and media discourse than in the two other countries examined here. The explanation could be as simple as the outright popularity of the war in Denmark, or at least the lack of clear misgivings about deploying and losing so many soldiers. That suggests the war can be viewed in Denmark as a political commodity, exploitable by the government in the country's decade-long domestic political disputes over defending Danish values and identity. It is indicative that in virtually all the political speeches on the war and its human cost, priority is given to the image of Danish soldiers travelling to Afghanistan in defence of Danish values. The need to help Afghans prevent terrorism and instability from spreading is also mentioned, but never with comparable emphasis.

The comparison between Denmark and Germany reveals the impact of political leadership on how dead soldiers are received and a war is perceived by society at large. The convention of bringing the dead of Afghanistan back home has shed light on the changed role of the soldier. During the Cold War, the tenet was to 'be able to fight in order to avoid it'. Today what's needed, in addition to the ability, is 'wanting to fight' (Bach 2009: 40). There is little new in such an attitude for Britain, but the consequences are momentous for Denmark and Germany, where the will to fight traditionally has been disparaged as machismo – a threat to a peace-loving, egalitarian society. It should be added that this role change seems to have caused fewer problems among the soldiers themselves than among the politicians. It is the political dimension that is different. In Britain and Denmark, the contributions of uniformed service personnel are openly recognised and lauded; in Germany, this is done timidly and reluctantly.

My book has shown that death on the field of battle is experienced much the same way by the men and women in uniform, and by the next-of-kin. The need to share the resulting feeling of loss shows no national differences. The method of sharing varies according to national forms of expression, but above all it is dependent on political will. In much contemporary writing on war and modern societies, attention is given to the complex and often contradictory impressions we are provided; that it is always difficult for the public to understand what is going on (Johnson and Tierney 2006, Hoskins and O'Loughlin 2010). To me, focusing on the dead and how they are honoured narrows down the complexities considerably. They are the true price of war.

Bibliography

Ainsworth, B. 2009. We Cannot Let These 100 Deaths Deflect Us. *The Times*, 9 December.

Allan, A. 2009. *Afghanistan: A Tour of Duty.* London: Third Millennium Publishing.

Allen, N. 2010. *EMBED. With the Worlds Armies in Afghanistan.* The Mill: Spellmount.

Allen, N. and Savill, R. 2008. Sarah Bryant's body returned from Afghanistan. *The Telegraph* [Online], 23 June 2008. Available at: http://www.telegraph.co.uk/news/worldnews/asia/afghanistan/2182404/Sarah-Bryants-body-returned-from-Afghanistan.html [accessed: 22 October 2011].

Allen, V. and Groves, J. 2009. Brown should resign, says father of soldier killed with commander Rupert Thorneloe. *Daily Mail* [Online], 2 November 2009. Available at: http://www.dailymail.co.uk/news/article-1224514/Father-soldier-killed-Rupert-Thorneloe-says-Gordon-Brown-liar-clueless.html [accessed: 6 June 2012].

Alvarez, M.R. and Brehm, J. 2002. *Hard Choices, Easy Answers.* Cornell: Cornell University Press.

Ancram, E.o. 2001. International terrorism. Hansard [Online], 12 December 2001, 376: 846–93. Available at: http://hansard.millbanksystems.com/commons/2001/dec/12/international-terrorism#S6CV0376P0_20011212_HOC_242.

ARD. 2010. Tagesschau, 8 August 2010. Available at: www.ard.de [accessed: 9 August 2010].

Ashdown, P. 2009. What we must do to win this war in Afghanistan. *The Independent*, 22 July.

Ashplant, T.G., Dawson, G. and Roper, M. 2004. The politics of war memory and commemoration. Contexts, structures and dynamics. In Ashplant, T.G., Dawson, G. and Roper, M. (eds) *Commemorating War.* New Brunswick: Transaction Publishers, 3–85.

Auswärtiges Amt, Bundesministerium des Inneren, Bundesministerium der Verteidigung and Bundesministerium für wirtschaftliche Zusammenarbeit und Entwicklung. 2008. *Das Afghanistan-Konzept der Bundesregierung.* Berlin: Die Bundesregierung.

Auswärtiges Amt. 2001. *Übereinkommen über vorläufige Regelungen in Afghanistan bis zur Wiederherstellung dauerhafter staatlicher Institutionen.* Berlin: Auswärtiges Amt.

Avisen.dk. 2012. Hærværksramt mindesten for militærnægtelse genopstilles 5. september i Gribskov [Online], 3 September 2012. Available at: http://avisen.dk

/blogs/arnehansen/haervaerksramt-mindesten-for-militaernaegtelse-genops_35774 [accessed: 12 December 2012].

Bach, A. 2009. Tod und Verwundung. *Das schwarze Barett*, 41: 39–41.

Bacia, H. 2008. Gestorben für eine friedlichere Zukunft in Afghanistan. *Frankfurter Allgemeine Zeitung* [Online], 1 September 2008. Available at: http://www.faz.net/aktuell/politik/ausland/trauergottesdienst-gestorben-fuer-eine-friedlichere-zukunft-in-afghanistan-1697782.html [accessed: 16 October 2011].

Bahr, E. 2000. Ein Protektorat wird selbständig. *Die Zeit*, 23: 6–7.

Bake, J. 2010. Das Bild vom demokratischen Soldaten: Erste Ergebnisse der empirischen Fallstudie zur Bundeswehr. In Dörfler-Dierken, A. and Kümmel, G. (eds) *Identität, Selbstverständnis, Berufsbild. Implikationen der neuen Einsatzrealität für die Bundeswehr.* Wiesbaden: VS-Verlag füs Sozialwissenschaften, 129–60.

Baker, P. 2010. Family fight for memorial honour. BBC News [Online]. Available at: http://news.bbc.co.uk/2/hi/uk_news/england/essex/8474288.stm [accessed: 19 February 2011].

Barnwell, C.A.B. 2010. June 2008. In Mcnab, A. (ed.) *SPOKEN FROM THE FRONT. Real Voices from the Battlefields of Afghanistan.* London: Corgi Books, 380–84.

Barrett, E. 2009. Queen sends message on first Armed Forces Day. *The Independent* [Online], 27 June 2009. Available at: http://www.independent.co.uk/news/uk/home-news/queen-sends-message-on-first-armed-forces-day-1721808.html [accessed: 17 August 2012].

Bates, C. 2010. The smile that defies the Taliban: Afghan teenager whose mutilated face shocked the world unveils her new image. *Daily Mail* [Online], 13 October 2010. Available at: http://www.dailymail.co.uk/news/article-1319804/Afghan-girl-Bibi-Aisha-nose-ears-hacked-Taliban-rule-gets-new-face.html#ixzz1d17LWIMd [accessed: 14 October 2010].

Bates, T.G. 2011. *The Reagan Rhetoric. History and memory in 1980s America.* Illinois: Northern Illinois University Press.

Baudrillard, J. and Patton, P. 1995. *The Gulf War did not take place.* Bloomington: Indiana University Press.

BBC. 2001. A liberating experience. BBC News [Online]. Available at: http://news.bbc.co.uk/2/hi/uk_news/1656237.stm [accessed: 15 February 2011].

BBC. 2002. Britain's al-Qaeda connections. BBC News [Online], 29 January. Available at: http://news.bbc.co.uk/2/hi/uk/1775683.stm [accessed: 16 October 2012].

BBC. 2006a. Afghanistan mission is 'cuckoo'. BBC News [Online], 29 October 2006. Available at: http://news.bbc.co.uk/2/hi/uk_news/politics/6096206.stm [accessed: 17 November 2011].

BBC. 2006b. UK troops 'to target terrorists'. BBC News [Online], 24 April 2006. Available at: http://news.bbc.co.uk/2/hi/uk_news/4935532.stm [accessed: 22 November 2011].

BBC. 2009a. British military deaths in Iraq. BBC News [Online]. Available at: http://www.bbc.co.uk/news/uk-10637526.

BBC. 2009b. British troops get delayed Merlin. BBC News [Online], 16 November 2009. Available at: http://news.bbc.co.uk/2/hi/uk_news/8361336.stm [accessed: 9 February 2012].

BBC. 2009c. Forces chief urges public to support Afghan mission. BBC News [Online], 3 December 2009. Available at: http://news.bbc.co.uk/2/hi/uk_news/8394269.stm [accessed: 17 October 2011].

BBC. 2009d. Thousands Attend Armed Forces Day. BBC News [Online], 27 June 2009. Available at: http://news.bbc.co.uk/2/hi/8121826.stm [accessed: 19 August 2012].

BBC. 2010a. Afghanistan death soldier, 18, on village war memorial. BBC News [Online], 11 November 2010. Available at: http://www.bbc.co.uk/news/uk-england-hereford-worcester-11732744 [accessed: 11 June 2012].

BBC. 2010b. UK death toll in Afghanistan conflict reaches 300. BBC News [Online], 21 June 2010. Available at: http://www.bbc.co.uk/news/10360705 [accessed: 13 March 2013].

BBC. 2012. Public opinion about the Afghan war changes. *BBC History*.

Beattie, D. 2009. *Task Force Helmand*. London: Simon & Schuster.

Bergmann, A. 2011. Darstellungen des Tötens und Getötetwerdens in der deutschen Berichterstattung über den Afghanistan- und Irakkrieg. *Paragrana*, 20(1): 293–315.

Berinsky, A.J. 2007. Assuming the Costs of War: Events, Elites, and American Public Support for Military Conflict. *The Journal of Politics*, 69(4): 975–97.

Berlin-gegen-Krieg.de. 2009. Berlin-gegen-Krieg. Available at: http://www.berlin-gegen-krieg.de/ [accessed: 17 February 2011].

Berliner Zeitung. 2001. Kabul gefallen, Taliban fliehen. *Berliner Zeitung* [Online], 14 November 2001. Available at: http://www.bz-berlin.de/archiv/kabul-gefallen-taliban-fliehen-article22210.html [accessed: 15 September 2012].

Berlingske. 2008. Regler hæmmer krigen mod Taleban. *Berlingske* [Online], 27 October 2008. Available at: http://www.b.dk/danmark/regler-haemmer-krigen-mod-taleban [accessed: 28 January 2011].

Berlingske. 2012. Afghansk succes eller fiasko. *Berlingske* [Online], 22 August 2012. Available at: http://www.b.dk/kommentarer/afghansk-succes-eller-fiasko [accessed: 25 August 2012].

Betts, R.K. 1999. Must War Find A Way? *International Security*, 24(2): 166–98.

Biddle, S.D. 2004. *Military Power. Explaining Victory and Defeat in Modern Battle*. Princeton, NJ: Princeton University Press.

Bieber, J. 2010. Cover photo. *Time Magazine*, 176(6).

Bild. 2006. Schock-Fotos von deutschen Soldaten *Bild* [Online], 26 October 2006. Available at: http://www.bild.de/news/aktuell/news/afghanistan-schaedel-soldaten-verhoer-964400.bild.html [accessed: 18 November 2011].

Bild. 2010. Trauerrede von Karl-Theodor zu Guttenberg: 'Ruht in Frieden, Soldaten, und seid in Gottes Segen'. *Bild* [Online], 10 April 2010. Available at: http://

www.bild.de/politik/2010/soldat/karl-theodor-zu-guttenberg-in-selsingen-12150944.bild.html# [accessed: 11 April 2010].

Bird, S. and Fairweather, C. 2009. IEDs and military fatalities in Iraq and Afghanistan. *RUSI Journal*, 154(4): 30–38.

Birkeland, S.A. 2010. *Krigens ansigt. Danske soldater i Afghanistan.* København: Gyldendal.

Birta, K. 2010. Ein Stück Einsatzrealität. *Bundeswehr aktuell*, 46(28): 3.

Bishop, P. 2008. *3 PARA.* London: Harper Press.

Bittner, J. 2007. Sterben wofür? *Die Zeit* [Online], 24 May 2007. Available at: http://www.zeit.de/2007/22/Afghanistan-Ehrenmal [accessed: 18 November 2012].

Blair, T. 1999. Doctrine of the International Community. *PBS Newshour* [Online], 22 April 1999. Available at: http://www.pbs.org/newshour/bb/international/jan-june99/blair_doctrine4-23.html [accessed: 13 March 2013].

Blair, T. 2001a. Coalition against International Terrorism. Hansard, 8 October, 372: 811–14.

Blair, T. 2001b. Full text of Tony Blair's speech to parliament. *The Guardian* [Online], 4 October. Available at: http://www.guardian.co.uk/world/2001/oct/04/september11.usa3 [accessed: 13 November 2011].

Blair, T. 2001c. The full text of Tony Blair's speech to the Welsh assembly. *The Guardian* [Online], 30 October. Available at: http://www.guardian.co.uk/world/2001/oct/30/september11.usa3?INTCMP=ILCNETTXT3487 [accessed: 12 November 2011].

Blair, T. 2001d. International Terrorism and Attacks in the USA. Hansard, 14 September, 603–7.

Blair, T. 2001e. Tony Blair's Afghanistan Speech. *Language in use* [Online], 7 October. Available at: http://www.putlearningfirst.com/language/20rhet/blair.html [accessed: 12 October 2012].

Blair, T. 2001f. Text: Tony Blair's statement. *The Guardian*, 7 October.

Blair, T. 2007. House of Commons Parliamentary Debates, 'Prime Minister's Engagements'. Hansard [Online], 27 June 2007, 323. Available at: http://www.publications.parliament.uk/pa/cm200607/cmhansrd/cm070627/debtext/70627-0002.htm [accessed: 16 March 2012].

Blair, T. 2010. *A Journey. My Political Life.* London: Hutchinson.

Blechschmidt, P. 2006a. Merkel muss weiter nein sagen. *Süddeutsche Zeitung*, 14 November 2006, 3.

Blechschmidt, P. 2006b. Wenn das Lächeln gefriert. *Süddeutsche Zeitung*, 2 June 2006.

Blechschmidt, P. 2008. Die Macht der Gewöhnung. *Süddeutsche Zeitung* [Online], 20 October 2008. Available at: http://www.sueddeutsche.de/politik/tote-deutsche-in-afghanistan-die-macht-der-gewoehnung-1.522648.

Blome, N. 2002. Die Deutschen und der Krieg. *Welt*, 8 March 2002.

Bloxham, A. 2011. William Hague: Talks with Taliban 'distasteful'. *The Telegraph* [Online], 23 June 2011. Available at: http://www.telegraph.co.uk/news/news

topics/onthefrontline/8593481/William-Hague-Talks-with-Taliban-distasteful.html [accessed: 24 June 2011].

bmvg. 1994. *Weissbuch 1994: Weissbuch zur Sicherheit der Bundesrepublik Deutschland und zur Lage und Zukunft der Bundeswehr.* Bonn: Bundesministerium der Verteidigung.

bmvg. 2009. *Das Ehrenmal der Bundeswehr. Den Toten unserer Bundeswehr – Für Frieden, Recht und Freiheit.* Berlin: Bundesministerium der Verteidigung.

Boas, K. 2007. 1864 lurer stadig i det danske baghoved. *Kristeligt Dagblad* [Online], 7 May 2007. Available at: http://www.kristeligt-dagblad.dk [accessed: 18 March 2011].

Bohannon, J. 2011. Counting the Dead in Afghanistan. *Science*, March, 331: 1256–60.

Book of Remembrance. 2008. *Book of Remembrance for those that gave their lives in Afghanistan.* London: Campaign Memorial Books.

Boorman, D. 2005. *A Century of Remembrance.* Barnsley: Pen & Sword.

Borg, A. 1991. *War Memorials. From Antiquity to the Present.* London: Leo Cooper.

Boulding, K.E. 1978. *Stable Peace.* Austin: University of Texas Press.

Boyce, S.M. 2002. Annual Chief of Defence Staff Lecture: UK Strategic Choices Following the Strategic Defence Review & 11th September. *RUSI Journal*, February, 1–7.

Braestrup, P. 1985. *Battle Lines. Report of the Twentieth Century Fund Task Force on the Military and the Media.* New York: Priority Press.

Bredal, B. 2011. Guds fred med vore døde. *Politiken* [Online], 17 January 2011. Available at: http://politiken.dk/debat/signatur/ECE1169015/guds-fred-med-vore-doede/ [accessed: 17 January 2011].

Brendle, F. 2007. 'Süß und ehrenvoll …' 2008 erhalten die 'fürs Vaterland gestorbenen' Soldaten ein Gedenkstätte. *Junge Welt*, 14 November 2007, 10.

British Legion. 2013. Tributes – How things began at Royal Wootton Bassett. *The Royal British Legion, The Wootton Basset Branch* [Online], January 2013. Available at: http://www.britishlegion.org.uk/branches/wootton-bassett/repatriation-information [accessed: 14 March 2013].

Brix, K. 2008. Taleban truer Danmark. *Ekstra Bladet* [Online], 20 October 2008. Available at: http://ekstrabladet.dk/nyheder/krigogkatastrofer/article1072352.ece [accessed: 11 November 2011].

Brix, K. and Cordsen, C. 2010. Stor opbakning til Søren Gade. *Jyllands-Posten* [Online], 12 February 2010. Available at: http://jyllands-posten.dk/indland/article1979265.ece [accessed: 22 May 2010].

Bromley, M. and Cushion, S. 2011. Media Fundamentalism. The immediate response of the UK national press to terrorism – from 9/11 to 7/7. In Zelizer, B. and Allan, S. (eds) *Journalism after September 11.* London: Routledge, 212–31.

Brooke, C. 2011. Wounded Ben's walk of courage: Veteran of Afghanistan leads tributes to those who made the ultimate sacrifice. *Daily Mail* [Online], 14 November 2011. Available at: http://www.dailymail.co.uk/news/article-

2061140/Remembrance-Sunday-2011-Veteran-Afghanistan-leads-tributes.html#ixzz1oEQ1a15L [accessed: 19 December 2011].

Brown, G. 2009. Gordon Brown on Remembrance Sunday [Online], 8 November 2009. Available at: http://www.labour.org.uk/gordon-brown-on-remembrance-sunday,2009-11-08 [accessed: 19 November 2011].

Browne, D. 2007. Des Browne's statement on troops for Afghanistan. *The Guardian*, 23 February.

Bruns, T. 2002. Landung in der Realität. *Welt* [Online], 30 January 2002. Available at: http://www.welt.de/print-welt/article371173/Landung-in-der-Realitaet.html [accessed: 28 May 2012].

Brück, T., de Groot, O. and Schneider, F. 2010. Ein Erste Schätzung der wirtschaftlichen Kosten der deutschen Beteiligung am Krieg in Afghanistan. *Wochenbericht des DIW Berlin*, 21.

Brøndum, C. 2006. Danskenes længste kamp. *Berlingske Tidende* [Online], 15 October 2006. Available at: http://www.b.dk/globalt/danskernes-laengste-kamp [accessed: 13 March 2013].

Brøndum, C. 2008a. Danskerne har vænnet seg til tab i Afghanistan. *Berlingske Tidende*, 26 August.

Brøndum, C. 2008b. Umuligt at tælle de dræbte. *Berlingske* [Online], 27 October 2008. Available at: http://www.b.dk/danmark/umuligt-taelle-de-draebte [accessed: 1 June 2010].

BT. 2007. Gade: Køretøj måtte ikke falde i fjendens hænder. BT [Online], 17 October 2007. Available at: http://www.bt.dk/danmark/gade-koeretoej-maatte-ikke-falde-i-fjendens-haender [accessed: 19 May 2011].

BT. 2008. Dræbte soldater lever videre i cyberspace. BT [Online], 1 April 2008. Available at: http://www.bt.dk/danmark/draebte-soldater-lever-videre-i-cyber space [accessed: 19 September 2011].

BT. 2011. Veteraner: Interner Hizb-ut-Tahrir. BT [Online], 19 January 2011. Available at: http://www.bt.dk/danmark/veteraner-interner-hizb-ut-tahrir-0 [accessed: 10 March 2013].

Buck, C. 2006. Nato-Partner setzen Deutschland unter Druck. *Welt* [Online], 18 November 2006. Available at: http://www.welt.de/print-welt/article95261/Nato-Partner-setzen-Deutschland-unter-Druck.html [accessed: 14 May 2011].

Bulmahn, T. 2008. Öffentliche Wahrnehmung der Bundeswehr. Medienberichte und Medienwirkungen. In Hettling, M. and Echterkamp, J. (eds) *Bedingt erinnerungsbereit. Soldatengedenken in der Bundesrepublik.* Göttingen: Vandenhoeck & Ruprecht, 151–61.

Bundesregierung. 2008. *Frieden und Entwicklung in Afghanistan – Sicherheit für uns.* Berlin: Die Bundesregierung.

Bundesregierung. 2010. Fortschrittsbericht Afghanistan zur Unterrichtung des Deutschen Bundestags. December, Berlin: Die Bundesregierung.

Bundestag. 2001. Stenographischer Bericht, 202. Sitzung. Plenarprotokoll. 16 November 2001. Berlin: Deutscher Bundestag.

Bundestag. 2007. Antrag der Abgeordneten Dr. Lukrezia Jochimsen, Dr. Norman Paech, Dr. Petra Sitte, Dr. Lothar Bisky, Volker Schneider (Saarbrücken), Dr. Gregor Gysi, Oskar Lafontaine und der Fraktion DIE LINKE. Ein Mahnmal – Kein Ehrenmal Gegen Kriege – Mahnmal für die Opfer der gegenwärtigen Kriege. 4 July 2007, Drucksache 16/5891. Berlin: Deutscher Bundestag.

Bundestag. 2009. Ehrung ehemaliger Wehrmachtssoldaten am Ehrenmal der Bundeswehr. 12 November 2009, Drucksache 17/32. Berlin: Deutscher Bundestag.

Bundestag. 2010a. Antwort der Bundesregierung auf die Kleine Anfrage der Abgeordneten Ulla Jelpke, Petra Pau, Wolfgang Gehrcke, weiterer Abgeordneter und der Fraktion DIE LINKE. – Drucksache 17/32 – Ehrung ehemaliger Wehrmachtssoldaten am Ehrenmal der Bundeswehr (Ergänzende Antwort zu Drucksache 17/199). 26 March 2010, Drucksache 17/1273. Berlin: Deutscher Bundestag.

Bundestag. 2010b. Folge 6: Gerhard Schröders Vertrauensfrage. *Serie Mistrauensvoten* [Online]. Available at: http://www.bundestag.de/dokumente/textarchiv/2010/32247430_misstrauensvotum06/index.jsp [accessed: 13 March 2013].

Bundestag. 2010c. Kleine Anfrage der Abgeordneten Ulla Jelpke, Dr. Lukrezia Jochimsen, Harald Koch, Jens Petermann und der Fraktion DIE LINKE. Erfahrungen mit dem Ehrenmal der Bundeswehr. 23 August 2010, Drucksache 17/2796. Berlin: Deutscher Bundestag.

Bundestag. 2010d. Kontingent aufgestockt. *Deutscher Bundestag* [Online], 26 February 2010. Available at: http://www.bundestag.de/dokumente/textarchiv/2010/28780774_kw08_de_afghanistan/index.html [accessed: 17 April 2012].

Bundestag. 2011. Wehrbeauftragter: Es gibt deutliche Verbesserungen. *Deutscher Bundestag* [Online]. Available at: http://www.bundestag.de/dokumente/textarchiv/2011/34442293_kw20_wehrbeauftragter/index.html.

Bundestag, D. 2001. Beratung des Antrags des Bundesregierung. Beteiligung bewaffneter deutscher Streitkräfte and dem Einsatz einer Internationalen Sicherheitsstützungstruppe in Afghanistan auf der Grundlage der Resolution 1386 (2001), 1383 (2001) and 1378 (2001) des Sicherheitsrates der Vereinten Nationen. 22 December, Berlin, 20821–58.

Bundeswehr. 2011a. Im Gedenken. Im Ehrenmal der Bundeswehr namentlich genannte verstorbene Bundeswehrangehörige, 2 June 2011. Available at: http://www.bundeswehr.de [accessed: 19 March 2012].

Bundeswehr. 2011b. Verband erkämpft bessere Einsatzversorgung. *Die Bundeswehr* [Online], 15 Oktober 2011, Pressemitteilung Nr. 31/2011.

Bundeswehr. 2012a. Einsatzzahlen – Die Stärke der deutschen Einsatzkontingente. 21 March 2012. Available at: http://www.bundeswehr.de [accessed: 2 April 2012].

Bundeswehr. 2012b. Todesfälle im Auslandseinsatz. bundeswehr.de [Online], 17 September 2012. Available at: http://www.bundeswehr.de [accessed: 15 March 2013].

Bush, G. 1992. US–Germany: Partners in leadership. *US Department of State Dispatch*, 4 May.

Butler, J. 2010. *Frames of War. When Is Life Grievable?* London: Verso.

Böhmer, D.-D. 2009. Die Deutschen müssen mehr Risiko eingehen. *Welt* [Online], 10 September 2009 [accessed: 24 March 2011].

Calder, A. 2004. *Disasters and Heroes. On War, Memory and Representation.* Cardiff: University of Wales Press.

Cameron, D. 2011. PM announces UK troop withdrawals from Afghanistan [Online], 6 July 2011. Available at: https://www.gov.uk/government/news/pm-announces-uk-troop-withdrawals-from-afghanistan [accessed: 14 March 2013].

Caruso, J.D. and Stougaard, K.E. 2009. Den sidste mission. *Berlingske Tidende* [Online], 25 April 2009. Available at: http://www.b.dk/krigen/den-sidste-mission [accessed: 22 June 2011].

Cathcart, B. 2008. An unprecedented focus on the wounded. *New Statesman*, 7 April 2008, 21.

Channel 4. 2009. Afghanistan War: Rupert Thorneloe's tribute to Mark Evison. *Channel 4 News* [Online], 5 October 2009. Available at: http://www.channel4.com/news/afghanistan-war-rupert-thorneloe-tribute-to-mark-evison [accessed: 8 May 2011].

Channel 4. 2010. Cameron in Afghanistan: this year is 'vital'. *Channel 4 News* [Online], 10 June 2010. Available at: http://www.channel4.com/news/articles/politics/domestic_politics/cameron%2Bin%2Bafghanistan%2Bthis%2Byear%2Bis%2Baposvitalapos/3676232.html [accessed: 17 February 2011].

Chapman, A.C. 2010. 12 September 2007 [diary]. In Mcnab, A. (ed.) *SPOKEN FROM THE FRONT. Real Voices from the Battlefields of Afghanistan.* London: Corgi Books, 330–31.

Chauvistré, E. 2001a. Operation 'Enduring War'. *die tageszeitung*, 27 September.

Chauvistré, E. 2001b. Überrüstet und überfordert. *die tageszeitung*, 29 October.

Chauvistré, E. 2009a. Robuste Illusionen: Nich schießen, wir sind Deutsche. *Interationale Politik*, March, 84–95.

Chauvistré, E. 2009b. *Wir Gutkrieger. Warum die Bundeswehr im Ausland scheitern wird.* Frankfurt: Campus Verlag.

Chin-A-Fo, H. 2010. Taliban happy to see Dutch leave. nrc.nl, 24 February.

Chivers, C.J. 2011. Vantage Point: The Challenges of Small-Unit Patrolling in Afghanistan. *New York Times* [Online], 13 January 2011. Available at: http://atwar.blogs.nytimes.com/2011/01/13/vantage-point-the-challenges-of-small-unit-patrolling-in-afghanistan/ [accessed: 15 January 2011].

Chrisafis, A. and Tilden, I. 2003. Pinter blasts 'Nazi America' and 'deluded idiot' Blair. *The Guardian* [Online], 11 June 2003 [accessed: 17 April 2011].

Christiansen, O.W. 2008. Mindesten for militærnægtere afsløret. Modkraft.dk [Online], 11 November 2008. Available at: http://spip.modkraft.dk/nyheder/article/mindesten-for-militaernaegtere [accessed: 1 December 2011].

Clarke, M. 2011. The Top Brass and the Politicians: Strained Relations. *RUSI.*

Coëpel, P. 1997. *Que maudite soit la guerre! Enquête sur un monument aux morts pacifiste.* Bricquebosq: Les Editions Des Champs.
Coker, C. 2001. *Humane Warfare.* London: Routledge.
College of Arms. 2011. *The College of Arms Newsletter* [Online], September 2011, 30. Available at: http://www.college-of-arms.gov.uk/Newsletter/030.pdf [accessed: 12 November 2011].
Collins, N. 2010. David Cameron: Britain 'paying high price' in Afghanistan. *The Telegraph*, 21 June.
Committee. 2012. 24 Journalists Killed in Afghanistan since 1992/Motive Confirmed Committee to Protect Journalists.
Connerton, P. 1989. *How Societies Remember.* Cambridge: Cambridge University Press.
Coppin, R. 1965. Remembrance Sunday. *Theology*, January, 525–30.
Cordesman, A.H. 2012. *Avoiding Creeping Defeat in Afghanistan: The Need for Realistic Assumptions, Strategy, and Plans.* Washington: CSIS.
Cordua, J. 2009. For venlig til at ville magten. *Information* [Online], 23 October 2009. Available at: http://www.information.dk/208280 [accessed: 10 May 2010].
Daily Mail. 2009. Off to a flyer: Thousands turn out to honour servicemen for Britain's first Armed Forces day. *Daily Mail* [Online], 28 June 2009. Available at: http://www.dailymail.co.uk/news/article-1195917/Off-flyer-Thousands-turn-honour-servicemen-Britains-Armed-Forces-day.html [accessed: 17 August 2012].
Dalager, S. 2010. Er krigen i Afghanistan ofrene værd? *Politiken* [Online], 18 March 2010. Available at: http://politiken.dk/debat/kroniker/ECE925105/er-krigen-i-afghanistan-ofrene-vaerd/ [accessed: 22 September 2011].
Dalgaard-Nielsen, A. 2008. *Umulig mission? Danmark i Afghanistan og Irak.* København: Gyldendal.
Damm, M.V. 2010. Værnepligten er sat ud af kraft. P3Nyheder [Online], 1 June 2010. Available at: http://www.dr.dk/P3/P3Nyheder/2010/06/01/062507.htm [accessed: 22 June 2012].
Danziger, D. 2010. *We Are Soldiers. Our Heroes. Their Stories. Real Life on the Front Line.* London: Sphere.
Dauber, C. 2001. Image as Argument: The Impact of Mogadishu on U.S. Military Intervention. *Armed Forces & Society*, 27(2): 205–29.
Davies, Q., Clark, B. and Sharp, M. 2008. *Report of Inquiry into National Recognition of our Armed Forces.* London: HMSO.
Davis, D. 2009a. Commons Debates. Hansard [Online], 2 November 2009, Column 575. Available at: http://www.publications.parliament.uk/pa/cm200809/cmhansrd/cm091102/debtext/91102-0004.htm [accessed: 11 June 2011].
Davis, J.W. 2009b. Ein Nationalfriedhof für deutsche Helden. *Süddeutsche Zeitung*, 7 September 2009, 2.
De Faldne, 2010. Directed by Heilbuth, P.-E. København: Danmarks Radio.
De hjemvendte. 2008. Directed by Lyster, H. København: DR2.
Defence Committee. 2009. *Helicopter Capability.* London: House of Commons.

DefenceFocus. 2011. In Memoriam. Deaths on Operations – 15 April 2011 to 19 May 2011. *DefenceFocus*, 252.

DeLillo, D. 1975. *End Zone*. New York: Pocket Books.

Demmer, U., Koelbl, S., Sandberg, B. and Szandar, A. 2006. Exzess am Hindukusch. *Der Spiegel*, 44.

Dempsey, J. 2006. Germany criticized for its training of Afghan police. *New York Times* [Online], 15 October 2006. Available at: http://www.nytimes.com/2006/11/15/world/europe/15iht-germany.3551337.html [accessed: 19 April 2011].

Denning, D. 2010. *In Pursuit of British Public Support for the Next War*. Norfolk, VA: National Defense University. Joint Forces Staff College. Joint Advanced Warfighting School.

Der Spiegel. 2006. Die Deutschen müssen das Töten lernen. *Der Spiegel*, 47.

Der Tagesspiegel. 2012. De Maizère will Ehrentag für Bundeswehr-Veteranen. *Der Tagesspiegel* [Online], 15 February 2012. Available at: http://www.tagesschau.de/inland/veteranentag104.html [accessed: 17 February 2012].

Deutsche Welle. 2007. Germans Divided on How to Honor Fallen Soldiers. *Deutsche Welle* [Online], 25 May 2007. Available at: http://www.dw-world.de/dw/article/0,,2557411,00.html [accessed: 19 November 2011].

Deutschlandfunk. 2011. Mehr als jeder vierte Rekrut schmeißt hin. Erste Zahlen über Abbrecherquoten bei der Bundeswehr. *Deutschlandfunk* [Online], 21 December 2011. Available at: http://www.dradio.de/aktuell/1634312/ [accessed: 3 January 2012].

Die Welt. 2011. Regierung kündigt Trendwende in Afghanistan an. *Die Welt* [Online], 11 January 2011. Available at: http://www.welt.de/politik/deutschland/article12098138/Regierung-kuendigt-Trendwende-in-Afghanistan-an.html [accessed: 11 January 2011].

Diner, D. 2002. *Feindbild Amerika. Über die Beständigkeit eines Ressentiments*. München: Propyläen.

Dirk. 2008. Deutsche Bundeswehr-Soldaten in Afghanistan fühlten sich angegriffen und feuerten in Kunduz: 3 tote Afghanen, 3 verletzte Afghanen, darunter 1 Kind. Soldatenglück.de [Online], 28 August 2008. Available at: http://soldatenglueck.de/2008/08/28/814/deutsche-bundeswehr-soldaten-in-afghanistan-fuehlten-sich-angegriffen-und-feuerten-in-kunduz-3-tote-afghanen-3-verletzte-darunter-1-kind/ [accessed: 19 November 2011].

Dirk. 2009a. Im Kopf geht der Krieg weiter – 'Angriff auf die Seele'. Soldatenglück.de [Online], 3 February 2009. Available at: http://soldatenglueck.de/2009/02/03/3494/im-kopf-geht-der-krieg-weiter-angriff-auf-die-seele/ [accessed: 19 March 2010].

Dirk. 2009b. Worte an und für Soldatinnen und Soldaten zum Volkstrauertag. Soldatenglück.de [Online], 15 November 2009. Available at: http://soldatenglueck.de/2009/11/15/24775/worte-an-und-fuer-die-soldatinnen-und-soldaten-zum-volkstrauertag/ [accessed: 19 May 2011].

Docherty, L. 2007. *Desert of Death. A Soldier's Journey from Iraq to Afghanistan*. London: Faber and Faber.

Dorman, A.M. 2009. *Blair's Successful War. British Military Intervention in Sierra Leone*. London: Ashgate.

Dowling, T. 1989. The Graves We Dug. *Army*, January, 36–39.

dpa. 2009. Linkspartei kritisiert neuen Tapferkeitsorden scharf. nachrichten.t-online.de [Online], 6 July 2009. Available at: http://nachrichten.t-online.de/linkspartei-kritisiert-neuen-tapferkeitsorden-scharf/id_19352804/index [accessed: 10 November 2010].

DR. 2008. USA og Danmark historisk tæt. dr.dk [Online], 28 February 2008. Available at: http://www.dr.dk/Nyheder/Politik/2008/02/28/101856.htm [accessed: 12 June 2011].

DR.Nyheder. 2007. Statsministeren vil have DR undersøgt. DR.Nyheder [Online], 25 January 2007. Available at: http://www.dr.dk/Nyheder/Politik/2007/01/25/064755.htm [accessed: 23 November 2011].

Drury, I. 2012. Taliban commander admits Islamist fighters can NOT win the war in Afghanistan as bloody stalemate continues. *Daily Mail* [Online], 11 July 2012. Available at: http://www.dailymail.co.uk/news/article-2172087/Taliban-commander-admits-Islamist-fighters-NOT-win-war-Afghanistan-bloody-stalemate-continues.html#ixzz2Lc88N4X2 [accessed: 15 September 2012].

Drury, I. and Wilkes, D. 2010. The Afghan war and a tale of two images: A near-empty Commons during war debate as tears flow at repatriation ceremony. *Daily Mail*, 10 September.

Dueholm, P. 2010. Danmark i Afghanistan – Hvorfor, hvordan og hvor længe? København: Udenrigsministeriet og Forsvarsministeriet.

Duffin, C. 2012. Save our war memorials: Heritage organisations form action group. *The Telegraph* [Online], 12 February 2012. Available at: http://www.telegraph.co.uk/history/britain-at-war/9076425/Save-our-war-memorials-heritage-organisations-form-action-group.html [accessed: 14 February 2012].

Dummet, M. 2010. Ministers 'united' on Afghanistan. BBC News, 22 May.

Dunn, T.N. and Larcombe, D. 2010. We're in for as long as it takes. *The Sun*, 8 November.

Dunne, P.J. and Nikolaidou, E.2011. Defence Spending and Economic Growth in the EU15. Carecon.org.

Dutter, B. 2001. Omar's fighters flee with their guns. *The Telegraph* [Online], 8 December 2001. Available at: http://www.telegraph.co.uk/news/worldnews/asia/afghanistan/1364716/Omars-fighters-flee-with-their-guns.html [accessed: 15 September 2011].

Echternkamp, J. 2008. Kein stilles Gedenken. Die Toten der Wehrmacht im Erinnerungskonflikt der Bundesrepublik. In Hettling, M. and Echternkamp, J. (eds) *Bedingt erinnerungsbereit. Soldatengedenken in der Bundesrepublik*. Göttingen: Vandenhoeck & Ruprecht, 46–57.

Eichenberg, R.C. 2005. Victory Has Many Friends. U.S. Public Opinion and the Use of Military Force, 1981–2005. *International Security*, 30(1): 140–77.

Eichmann, B. 1985. *Versteinert, verharmlost, vergessen: KZ-Gedenkstätten in der Bundesrepublik Deutschland.* Frankfurt: Fischer.

Eisele, A. 2005. Embed Cavallaro sees war from the inside. *The Hill*, 31 March.

Ekstrabladet. 2007. Mission umulig. Endnu et liv spilt. *Ekstrabladet*, 17 October 2007, 1.

Ellemann-Jensen, U. 2004. *Fodfejl: da Danmark svigtede under den kolde krig.* København: Gyldendal.

Elliot, L. 2009. War on the cheap is costing soldiers' lives. *The Guardian*, 13 July.

Entman, R.M. 1993. Framing: Toward clarification of a fractured paradigm. *Journal of Communication*, 43(4): 51–58.

Entman, R.M. 2004. *Projections of Power. Framing News, Public Opinion, and U.S. Foreign Policy.* Chicago: University of Chicago Press.

Erhart, H.-G. and Kaestner, R. 2008. *Afghanistan: Scheitern oder Strategiewechsel.* Hamburg: Hamburger Informationen zur Friedensforschung und Sicherheitspolitik.

Etherington, S.D. 2006. The Last Stand of the 44th Regiment – Gandamack, Afghanistan 13 Jan 1842. *British Army Review*, 140: 58–65.

Faber, K. 2010. Udskælt monument kan ikke stoppes. *Politiken* [Online], 2 November 2010. Available at: http://politiken.dk/ibyen/nyheder/gadeplan/ECE1099095/udskaeldt-monument-kan-ikke-stoppes/ [accessed: 13 March 2013].

Farrell, S. 2010. Embedistan. *The New York Times* [Online], 25 June 2010. Available at: http://atwar.blogs.nytimes.com/2010/06/25/embedistan-2/ [accessed: 25 June 2010].

Farrell, T. 2005. *The Norms of War. Cultural Beliefs and Modern Conflict.* London: Lynne Rienner Publishers.

FAZ. 2001. Höchstens tausend Soldaten. Berlin will keine Führungsrolle in Afghanistan-Schutztruppe. *Frankfurter Allgemeine Zeitung* [Online], 13 December 2001. Available at: http://www.seiten.faz-archiv.de/faz/20011212/fd3200112121208960.html [accessed: 2 May 2011].

FAZ. 2004. Des Kanzlers D-Day. *Frankfurter Allgemeine Zeitung* [Online], 7 June 2004. Available at: http://www.faz.net/artikel/C30089/kommentar-des-kanzlers-d-day-30113099.html [accessed: 19 March 2011].

FAZ. 2006. Das Bild des Bürgers in Uniform. *Frankfurter Allgemeine Zeitung* [Online], 26 October 2006. Available at: http://www.faz.net/artikel/C31013/soldatenfotos-das-bild-des-buergers-in-uniform-30095125.html [accessed: 7 June 2011].

FAZ. 2007. Jung präsentiert Entwurf für Ehrenmal. *Frankfurter Allgemeine Zeitung* [Online], 13 June 2007. Available at: http://www.faz.net/aktuell/politik/inland/bundeswehr-jung-praesentiert-entwurf-fuer-ehrenmal-1434746.html [accessed: 13 March 2013].

FAZ. 2008. Bundeswehrverband: Krieg in Afghanistan. *Frankfurter Allgemeine Zeitung*, 4 September 2008, 1.

FAZ. 2010. 'Deutschland verneigt sich mit Dank und Hochachtung' Merkel würdigt die in Afghanistan gefallenen Soldaten / Trauerfeier in Selsingen *Frankfurter*

Allgemeine Zeitung [Online], 10 April 2010. Available at: http://www.seiten.faz-archiv.de/faz/20100410/fd1n201004102662524.html [accessed: 11 April 2010].

FCO and DFID. 2005. *The UK and Afghanistan*. London: Foreign and Commonwealth Office and The Department for International Development.

Feaver, P.D. and Gelpi, C. 2004. *Choosing Your Battles: American Civil-Military Relations And The Use Of Force*. Princeton: Princeton University Press.

Feaver, P.D., Kohn, R.H. and Cohn, L.P. 2001. The Gap Between the Military and Civilian in the United States in Perspective. In Feaver, P.D. and Kohn, R.H. (eds) *Soldiers and Civilians. The Civil-Military Gap and American National Security*. Cambridge, Massachusetts: MIT Press, 1–11.

Feldmeyer, K. 2001. Die Bundeswehr fordert ein robustes Mandat. *Frankfurter Allgemeine Zeitung*, 17 December 2001.

Fiebig, R. and Pietsch, C. 2010. Der Bundeswehr-Einsatz in Afghanistan. *Europäische Sicherheit*, 59(5): 24–7.

Fischer, J. 1999. Rede des Außenministers zum Natoeinsatz im Kosovo. *Archiv Grünes Gedächtnis*. Heinrich Böll Stiftung.

Fischer, M. 2010. Taliban greifen Bundeswehr an. n-tv [Online], 3 April 2010. Available at: http://www.n-tv.de/politik/dossier/Taliban-greifen-Bundeswehr-an-article807583.html [accessed: 4 April 2010].

Focus. 2010. Zitate: Vom 'Stabilisierungseinsatz' zum 'Krieg' [Online], 22 April 2010. Available at: http://www.focus.de/politik/ausland/konflikte-zitate-vom-und132stabilisierungseinsatzund147-zum-und132kriegund147_aid_501164.html [accessed: 19 August 2011].

Focus. 2011. Mehrere Hundert freiwillige Rekruten geben auf [Online], 1 August 2011. Available at: http://www.focus.de/politik/deutschland/bundeswehr-mehrere-hundert-freiwillige-rekruten-geben-auf_aid_651584.html [accessed: 22 August 2011].

Folk & Forsvar. 2009. Flagdag 5. september: Arrangementer, *Folk & Forsvar og Danske Soldater- og Marineforeningers Fællesråd* [Online], August 2009. Available at: http://www.beredskab.dk/fileadmin/downloads/Flagdag_2009.pdf [accessed: 19 November 2011].

Fonseca, D. 2009. Facebook livsfarligt for danske soldater. *Ekstra Bladet* [Online], 25 March 2009. Available at: http://ekstrabladet.dk/nyheder/samfund/article1144906.ece [accessed: 22 February 2011].

Foreign Affairs Committee. 2011. *The UK's foreign policy approach to Afghanistan and Pakistan*. London: House of Commons.

Forsvaret. 2006. Bilag til forsvarsministerens skrivelse af 21. September 2006.

Forsvaret. 2012. Opfølgningsundersøgelsen – positive aspekter. *Forsvarets Psykologer* [Online], 2012. Available at: https://imp.fak.dk/CMSPages/kategoriindhold.aspx?kat=Opf%C3%B8lgningsunders%C3%B8gelsen%20-%20positive%20aspekter [accessed: 16 November 2012].

Forsvaret. 2013. Liste udvisende navne på omkomne på Monumentet over Danmarks Internationale Indsats siden 1948. *Forsvaret* [Online], 1 February

2013. Available at: http://veteran.forsvaret.dk/SiteCollectionDocuments/liste%20over%20faldne%202013-02.pdf [accessed: 14 March 2013].

Forsvarsministeriet. 2006. Redegørelse for faktuelle og retlige spørgsmål i tilknytning til danske styrkers tilbageholdelse og overladelse af personer i Afghanistan i 1. halvår 2002, 13 December 2006. Available at: http://www.dr.dk/forbrug/portal/cms-assets/hemmeligkrig/Redegørelse_13122006.pdf [accessed: 7 June 2011].

Forsvarsministeriet. 2011. The monument to Denmark's international effort. [Online], 11 March 2011. Available at: http://www.fmn.dk/eng/allabout/Pages/ThemonumenttoDenmarksinternationaleffort.aspx [accessed: 16 March 2011].

Forsvarsministeriet. 2012. Fundamentet til det danske kursskifte i Afghanistan er færdigstøbt 8 May 2012. Available at: http://www.fmn.dk/nyheder/Arkiv/Afghanistan/Pages/Fundamentettildetdanskekursskifteerfaerdigstoebt.aspx [accessed: 9 May 2012].

Frankfurter Rundschau. 2009a. Deutschland im Kreuzfeuer der Kritik. *Frankfurter Rundschau* [Online], 7 September 2009. Available at: http://www.fr-online.de/einsatz-in-afghanistan/afghanistan-einsatz-deutschland-im-kreuzfeuer-der-kritik,1477334,2819476.html [accessed: 17 March 2011].

Frankfurter Rundschau. 2009b. Ruf nach Merkel. Debatte um Afghanistan-Einsatz hebt an. *Frankfurter Rundschau* [Online], 7 September 2009. Available at: http://www.fr-online.de/politik/debatte-um-afghanistan-einsatz-ruf-nach-merkel,1472596,3329484.html [accessed: 19 November 2011].

Frankfurter Rundschau. 2010a. Ruf nach besserer Ausrüstung. *Frankfurter Rundschau*, 6 April 2010, 3.

Frankfurter Rundschau. 2010b. Trauerfeier für tote Soldaten verschoben. *Frankfurter Rundschau* [Online], 18 April 2010. Available at: http://www.fr-online.de/politik/bundeswehr-trauerfeier-fuer-tote-soldaten-verschoben,1472596,2678470.html [accessed: 22 April 2010].

fred.dk. 2008. Mindesten for militærnægtelse [Online], November 2008. Available at: http://fred.dk/amk/aktu1-08.htm [accessed: 1 December 2011].

Freuding, C. 2009. Die *Operation Enduring Freedom* und *Active Endeavour* – Deutschlands militärischer Beitrag zum kampf gegen den internationalen Terrorismus. In Gießmann, H.J. and Wagner, A. (eds) *Armee im Einsatz*. Baden-Baden: Nomos, 340–52.

Freund, C. 2008. *Antiamerikanismus in Deutschland: Die Veränderung der Wahrnehmung Amerikas aus deutscher Sicht im 20. Jahrhundert und zu Beginn des neuen Jahrtausends*: Grin Verlag.

Fried, N. 2006. Der Süden bleibt tabu. *Süddeutsche Zeitung*, 14 November 2006, 3.

Furhmeister, C. 2009. Skandalöse Stille, klammheimliche Änderungen. Bemerkungen zum geplanten Ehrenmal der Bundeswehr. In Hauswedell, C. (ed.) *Soldatentod in heutigen Kriegen. Herausforderungen für politische Normenbildung und Erinnerungskultur*. Rehberg-Loccum: Loccummer Protokolle, 149–59.

Furness, H. 2012. 'Poppy fascism' row reignites over accusations appeal has been hijacked by politicians and 'B-list celebrities'. *The Telegraph* [Online],

24 October 2012. Available at: http://www.telegraph.co.uk/news/uknews/defence/9629864/Poppy-fascism-row-reignites-over-accusations-appeal-has-been-hijacked-by-politicians-and-B-list-celebrities.html [accessed: 25 October 2012].

Gade, S. 2009. Farvel til våbnene? *Jyllands-Posten* [Online], 28 December 2009. Available at: http://jyllands-posten.dk/opinion/kronik/article4549487.ece [accessed: 16 March 2011].

Gade, S. and Tørnæs, U. 2006. Det lange sæje træk i Afghanistan. *Jyllands-Posten* [Online], 29 September 2006 [accessed: 17 January 2012].

Garderportal. 2008. Mindeside for overkonstabel Christian Raaschou. garderportal.dk [Online], April 2008. Available at: http://www.garderportal.dk/MindesideOverkonstabelChristianRaaschou.html [accessed: 19 September 2011].

Garderportal. 2009. Mindeside: Major Anders Stæhr Storrud. *garderportal.dk* [Online], Spring 2009. Available at: http://www.garderportal.dk/MindesideMajorAndersStaehrStorrud.html [accessed: 22 June 2011].

Gebauer, M. 2010. Merkel trauert öffentlich – nun doch. *Der Spiegel* [Online], 8 April 2010. Available at: http://www.spiegel.de/politik/ausland/zeremonie-fuer-tote-soldaten-merkel-trauert-oeffentlich-nun-doch-a-687862.html [accessed: 9 April 2010].

Gebauer, M., Koelbl, S. and Reuter, C. 2011. Die Stimmung wird kippen. *Der Spiegel* [Online], 6 June 2011. Available at: http://www.spiegel.de/spiegel/print/d-78832431.html [accessed: 9 July 2011].

Gebauer, M. and Najafizada, S. 2009. Afghanen und Bundeswehr starten Großoperation gegen Taliban. *Der Spiegel* [Online], 20 July 2009. Available at: http://www.spiegel.de/politik/ausland/0,1518,637197,00.html [accessed: 3 May 2011].

Gelb, L.H. 2011. Joe Biden On Iraq, Iran, China and the Taliban. *Newsweek* [Online], 19 December 2011. Available at: http://www.thedailybeast.com/newsweek/2011/12/18/joe-biden-on-iraq-iran-china-and-the-taliban.html [accessed: 22 December 2011].

Gelpi, C., Feaver, P.D. and Reifler, J. 2009. *Paying the Human Costs of War: American Public Opinion and Casualties in Military Conflicts*. Princeton: Princeton University Press.

Gerges, D. 2012. 'A true hero and a shining star': Family pay tribute to RAF airman shot dead in Afghanistan. *Daily Mail*, 14 February.

Gilens, M. 2001. Political Ignorance and Collective Policy Preferences. *The American Political Science Review*, 95(2): 379–96.

Glover, J. 2009. Survey of public opinion on Afghan conflict finds support, and doubt. *The Guardian*, 13 July.

Gordon, A.I.C. 2002. Review of State Ceremonial 1996–2002. *British Army Review*, 130: 70–73.

Gram, S., Rasmussen, J.E. and Stenstrup, M. 2002. Mineryddere til Afghanistan. *Jyllands-Posten* [Online], 9 January. Available at: http://jp.dk/indland/article 608856.ece [accessed: 13 March 2013].

Grassmann, P. 2002. Deutsche Soldaten kämpfen an vorderster Front. *Süddeutsche Zeitung*, 9 March 2002, 37.

Graves, R. 1930. *But It Still Goes On: An Accumulation.* London: Jonathan Cape.

Graw, A. 2008. Jung will von einem Kriegszustand nichts wissen. *Welt online* [Online], 4 September 2008. Available at: http://www.welt.de/politik/article 2391658/Jung-will-von-einem-Kriegszustand-nichts-wissen.html [accessed: 16 October 2011].

Graw, A. 2010. Biden attackiert Deutschlands Rolle in Afghanistan. *Welt* [Online], 1 December 2011. Available at: http://www.welt.de/politik/ausland/article11 335869/Biden-attackiert-Deutschlands-Rolle-in-Afghanistan.html [accessed: 22 May 2011].

Graw, A. 2012. Die USA wollten zu viel, der Krieg ist verloren. *Welt* [Online], 13 March 2012. Available at: http://www.welt.de/debatte/article13919325/Die-USA-wollten-zu-viel-der-Krieg-ist-verloren.html [accessed: 15 March 2012].

Greenberg, A. 1989. Luyten's Cenotaph. *Journal of the Society of Architectural Historians*, 48: 5–23.

Grice, E. 2012. The widow of Falklands hero Colonel 'H' Jones says: 'I didn't need therapy – I had a close family'. *The Telegraph* [Online], 2 April 2012. Available at: http://www.telegraph.co.uk/news/worldnews/southamerica/falk landislands/9177481/The-widow-of-Falklands-hero-Colonel-H-Jones-says-I-didnt-need-therapy-I-had-a-close-family.html.

Griggs, I. 2007. Falklands War 25 years on: Tumbledown revisited. *The Independent*, 28 January.

Griswold, W. 1987. A Methodological Framework for the Sociology of Culture. In Clogg, C. (ed.) *Sociological Methodology.* Washington DC: American Sociological Association, 1–35.

Groos, H. 2011a. *'Das ist auch euer Krieg!' Deutsche Soldaten berichten von ihren Einsätze.* Frankfurt: Fischer Verlag.

Groos, H. 2011b. *Ein schöner Tag zum sterben. Als Bundeswehrärztin in Afghanistan.* Frankfurt am Main: Fischer Verlag.

Gross, E.E. 2009. *The Europeanization of National Foreign Policy: Continuity and Change in European Crisis Management.* Basingstoke: Palgrave Macmillan.

Grænseforeningen. 2010. Dybbølmærket. *Grænseforeningen* [Online], 2010. Available at: http://www.graenseforeningen.dk/leksikon/d/all/4579 [accessed: 15 November 2011].

Grøndahl, M., Rasmussen, T.R. and Sinclair, K. 2003. *Hizb ut-Tahrir i Danmark – Farlig fundamentalisme eller uskyldigt ungdomsoprør?* Århus: Århus Universitetsforlag.

Güßgen, F. 2007. SPD-Politikerin will Tote benennen. Stern.de [Online], 8 February 2007. Available at: http://www.stern.de/politik/deutschland/ehrenmal-debatte-spd-politikerin-will-tote-benennen-582201.html [accessed: 19 April 2011].

Haftendorn, H. 2002. One Year After 9/11: A Critical Appraisal of German-American Relations. *The Thyssen German American Dialogue Seminar Series* [Online], December 2002. Available at: http://www.sscnet.ucla.edu/polisci/faculty/trachtenberg/useur/haftendorn.pdf [accessed: 13 March 20013].

Halbwachs, M. 1997a. *La mémoire collective.* Paris: Albin Michel.

Halbwachs, M. 1997b. Mémoire collective et mémoire historique. *Le mémoire collective.* Paris: Albin Michel, 97–147.

Hammer, P. 2008. *Berichterstattung nach den Terroranschlägen vom 11. September- Ethik und Funktionen des Journalismus auf dem Prüfstand: Eine kommunikationswissenschaftliche Analyse und ein Essay.* München: GRIN Verlag.

Hammerstein, K.v., Koelbl, S., Szandar, A. and Yousafzai, S. 2008. K wie Krieg. *Spiegel* [Online], 8 September 2008, 37. Available at: http://www.spiegel.de/spiegel/print/d-59889996.html [accessed: 20 August 2011].

Hammond, P. 2012. Afghanistan: we owe it to all those who have sacrificed their lives to see this mission succesfully concluded. *The Telegraph*, 7 March.

Handelsblatt. 2008. Merkel kommt zum Bundeswehr-Gelöbnis. *Handelsblatt* [Online], 19 July 2008. Available at: http://www.handelsblatt.com/politik/deutschland/merkel-kommt-zum-bundeswehr-geloebnis/2991352.html [accessed: 2 September 2011].

Handelsblatt. 2010. Merkel erweist gefallenen Soldaten die letzte Ehre. *Handelsblatt* [Online], 8 April 2010. Available at: http://www.handelsblatt.com/politik/deutschland/nach-kritik-merkel-erweist-gefallenen-soldaten-die-letzte-ehre-seite-all/3407172-a [accessed: 9 April 2010].

Handelsblatt. 2011. Wehrbeauftragter kritisiert Einsparungen beim Afghanistan-Einsatz. *Handelsblatt* [Online], 6 June 2011. Available at: http://www.handelsblatt.com/politik/deutschland/handelsblatt-exklusiv-wehrbeauftragter-kritisiert-einsparungen-beim-afghanistan-einsatz/4257700.html [accessed: 7 June 2011].

Hansard. 2009. Commons debates [Online], 8 July 2009, Column 956. Available at: http://www.publications.parliament.uk/pa/cm200809/cmhansrd/cm090708/debtext/90708-0002.htm [accessed: 13 March 2013].

Hansard. 2011. Commons debates [Online], 10 November 2011, Column 472. Available at: http://www.publications.parliament.uk/pa/cm201011/cmhansrd/cm111110/debtext/111110-0002.htm [accessed: 12 November 2011].

Hansen, J.L. 2009a. Flagdag vigtig når granater falder. *TV2 Nyhederne* [Online], 5 September 2009. Available at: http://nyhederne.tv2.dk/article.php/id-24843758:flagdag-vigtig-n%C3%A5r-granater-falder.html [accessed: 17 March 2011].

Hansen, J.L. 2009b. Vil flage på halv på flagdag. *TV2 Nyhederne* [Online], 4 September 2009. Available at: http://nyhederne.tv2.dk/article.php/id-24824382:vil-flage-p%C3%A5-halv-p%C3%A5-flagdag.html [accessed: 19 May 2011].

Hansen, L. 2011. The politics of securitization and the Muhammad cartoon crisis: A post-structuralist perspective. *Security Dialogue*, 42(4–5): 357–69.

Harder, T. 2010. Jubler over billetsalg til Armadillo. *Ekstra Bladet* [Online], 28 May 2010 [accessed: 1 June 2010].

Harding, T. and Swaine, J. 2008. Afghanistan: British suffer four times as many casualties as Americans. *The Telegraph*, 21 December.

Harlow Council. 2010. Harlow War Memorial – Can you help? [Online], 17 June 2010. Available at: http://www.harlow.gov.uk/news/campaigns/harlow_war_memorial.aspx? [accessed: 17 February 2011].

Harnden, T. 2011. *Dead men risen: The Welsh Guards and the real story of Britain's war in Afghanistan.* London: Quercus.

Harrison, T. 2012. Wear your poppy with peaceful intent. *The Guardian* [Online], 25 October 2012. Available at: http://www.guardian.co.uk/commentisfree/2012/oct/25/poppy-remembrance-peace-glorify-war [accessed: 26 October 2012].

Hautkapp, D. 2010. Merkel und der Tod der Soldaten – Anfang der Trauerkultur. *Westdeutsche Zeitung* [Online], 9 April 2010. Available at: http://www.presseportal.de/pm/55903/1592592/waz-merkel-und-der-tod-der-soldaten-der-anfang-der-trauerkultur-leitartikel-von-dirk-hautkapp [accessed: 7 May 2010].

Heckmann, D.-O. 2008. 'Bestenfalls hat man ein freundliches Desinteresse'. General a.D. Reinhardt zum historischen Gelöbnis vor dem Reichstag. *Deutschlandfunk* [Online], 21 July 2008. Available at: http://www.dradio.de/dlf/sendungen/interview_dlf/819596/ [accessed: 18 November 2011].

Henken, L. 2006. Für uneingeschränkte Angriffsfähigkeit der Bundeswehr: Entwurf aus dem Verteidigungsministerium für ein 'Weißbuch 2006'. *IMI-Analyse*, 1–8.

Hepp, M. and Otto, V. 1996. *Soldaten sind Mörder. Dokumentation einer Debatte.* Berlin: Ch. Links Verlag.

Hettling, M. 2008. Militärisches Totengedenken in der Berliner Republik. Opfersemantik und politischer Auftrag. In Hettling, M. and Echterkamp, J. (eds) *Bedingt erinnerungsbereit. Soldatengedenken in der Bundesrepublik.* Göttingen: Vandenhoeck & Ruprecht, 11–21.

Hippler, J. 2012. Meinung: Afghanistan – wie weiter? *Bundeszentrale für politische Bildung – Newsletter*, 19 July.

Hoffman, C. 2012. Taliban Talks: Not the Solution to Afghanistan's Problems. *Kings of War* [Online], 12 March 2012. Available at: http://kingsofwar.org.uk/2012/03/taliban-talks-not-the-solution-to-afghanistans-problems/comment-page-1/ [accessed: 15 March 2012].

Hoffmann, T. 2010. Derfor fejlede dokumentaren Den Hemmelige Krig. Videnskap.dk [Online], 13 January 2010. Available at: http://videnskab.dk/kultur-samfund/derfor-fejlede-dokumentaren-den-hemmelige-krig [accessed: 14 January 2010].

Homberger, E. 1976. The story of the Cenotaph. *The Times Literary Supplement*, 12 November 1976, 1429–30.

Honeywell, D. 2008. Thousands call for National Remembrance Holiday to Honour the Fallen. *The Press* [Online], 25 July 2008. Available at: http://mobile.yorkpress.co.uk/news/3547589.THOUSANDS_CALL_FOR_NATIONAL_

REMEMBRANCE_HOLIDAY_TO_HONOUR_THE_FALLEN/ [accessed: 18 April 2011].

Hoskins, A. and O'Loughlin, B. 2010. *War and Media. The Emergence of Diffused War.* Cambridge: Polity Press.

House of Commons Defence Committee. 3 July 2007. UK operations in Afghanistan. Thirteenth Report of Session 2006–07. Report, together with formal minutes, oral and written evidence. London: House of Commons Defence Committee.

Howard, M. 1982/83. Reassurance and Deterrence: Western Defence in the 1980s. *Foreign Affairs*, 61(2): 309–24.

Hoyer, W. 2007. Plenardebatte 15 November 2007, 16/126. Berlin: Deutscher Bundestag, 13091–2.

Human Rights Watch. 2008. *'Troops in Contact' Airstrikes and Civilian Deaths in Afghanistan.* New York: Human Rights Watch.

Hundevadt, K. 2008. *I morgen angriber vi igen. Danmarks krig i Afghanistan.* København: Jyllands-Postens Forlag.

Hvass, J. 2013. Villy Søvndal om Afghanistan-plan: Man kan diskutere, hvilken stat vi forlader. *Politiken* [Online], 29 January 2013. Available at: http://politiken.dk/politik/ECE1882280/villy-soevndal-om-afghanistan-plan-man-kan-diskutere-hvilken-stat-vi-forlader/ [accessed: 2 February 2013].

Hyde, C. 2000. Casualty Aversion: Implications for Policy Makers and Senior Military Officers. *AerospacePower Journal*, 14(2): 17–27.

Hæren. 2007. Militær bisættelse af major Anders Storrud. *Hæren. Den Kongelige Livgarde* [Online], 24 October 2007. Available at: http://forsvaret.dk/LG/Nyt%20og%20Presse/Nyheder/Pages/Milit%C3%A6r%20bis%C3%A6ttelse%20af%20major%20Anders%20Storrud.aspx [accessed: 22 June 2011].

Hæren. 2008a. Mindeord og navnene på de tre dræbte soldater. *Hærens Operative Kommando* [Online], 20 December 2008. Available at: http://forsvaret.dk/HOK/Nyt%20og%20Presse/ISAF/Pages/Mindeordognavnenep%C3%A5detredr%C3%A6btesoldater.aspx [accessed: 11 November 2012].

Hæren. 2008b. SKIBET KIRKE: Mindedag for faldne soldater. *Hæren. Den Kongelige Livgarde* [Online], 17 November 2008. Available at: http://forsvaret.dk/LG/Nyt%20og%20Presse/Nyheder/Pages/Remembrance--Sunday.aspx [accessed: 18 August 2011].

Hæren. 2009. Mindeord. *Hærens Operative Kommando* [Online], 18 June 2009. Available at: http://forsvaret.dk/HOK/Nyt%20og%20Presse/ISAF/Pages/Mindeord-2.aspx [accessed: 20 August 2011].

Høy-Jensen, C. 2008. Hustru til dræbt soldat: De døde skal ikke på udstilling. *Politiken* [Online], 6 September 2008. Available at: http://politiken.dk/kultur/ECE563702/hustru-til-draebt-soldat-de-doede-skal-ikke-paa-udstilling/ [accessed: 22 June 2011].

Haass, R. 2009. *War of Necessity, War of Choice. A Memoir of Two Iraq Wars.* New York: Simon & Schuster.

iCasualties. 2013. Operation Enduring Freedom. Afghanistan. Fatalities By Year.

ICM. 2009. ICM poll on Afghanistan. UKPollingReport [Online], 13 July. Available at: http://ukpollingreport.co.uk/blog/archives/2198 [accessed: 11 March 2013].

Imperial War Museum. 2011–2012. In Our Own Words: Soldiers Thoughts from Afghanistan. *Imperial War Museum, Manchester* [Online], 1 October 2011 – 17 June 2012. Available at: http://www.iwm.org.uk/exhibitions/iwm-north/in-our-own-words-soldiers-thoughts-from-afghanistan [accessed: 15 January 2013].

Inacker, M. 2003. Eine Generalität ohne Zivilcourage. *Frankfurter Allgemeine Sonntagszeitung* [Online], 15 Juni 2003. Available at: http://www.seiten.faz-archiv.de/fas/20030615/sd1200306151907467.html [accessed: 16 March 2012].

Information. 2008. Soldater oprørte over flagfarce [Online], 17 December 2008. Available at: http://www.information.dk/telegram/177338 [accessed: 17 March 2011].

Information. 2010. 'Den Hemmelige Krig' endelig for retten [Online], 3 June 2010. Available at: http://www.information.dk/235140 [accessed: 23 October 2011].

Information. 2012a. Forsker: Krig kan ikke måles i kroner. Information.dk [Online], 16 January 2012 [accessed: 17 January 2012].

Information. 2012b. Forsvarminister: Krigen er prisen værd. Information.dk [Online], 16 January 2012. Available at: http://www.information.dk/telegram/290596 [accessed: 17 January 2012].

Information. 2012c. Krigsregnskab tæller ikke sårede soldater med. Information.dk [Online], 23 January 2012.

Information. 2012d. Soldater: Sårede skal hjælpes uanset prisen [Online], 23 January 2012. Available at: http://www.information.dk/telegram/291279 [accessed: 24 January 2012].

Information. 2012e. Søvndal forudser afghansk nedtur [Online], 13 March 2012. Available at: http://www.information.dk/print/295937 [accessed: 15 March 2012].

Inglis, K. 1992. War memorials: Ten Questions for Historians. *Guerres Mondiales et Conflicts Contemporains*, 167: 5–21.

ISAF. 2012. Troop numbers and contributions. *Afghanistan International Security Assistance Force* [Online], 2012. Available at: http://www.isaf.nato.int/troop-numbers-and-contributions/index.php [accessed: 1 February 2013].

Jaeger, F. 2009. Pathos der Opfer. *Der Tagesspiegel* [Online], 10 September 2009. Available at: http://www.tagesspiegel.de/kultur/ehrenmal-der-bundeswehr-pathos-der-opfer/1597188.html [accessed: 13 March 2013].

Jakobsen, P.V. 2004. Har Danmark et body bag syndrom? *Militært Tidsskrift*, 133(1): 94–114.

Jalland, P. 2010. *Death in War and Peace. Loss and Grief in England 1914–1970*. Oxford: Oxford University Press.

James, N. 2010. Interview with Steve McQueen. *Artslant*.

Jelpke, U. 2009. Bundeswehr trägt aktiv zur Verklärung der Wehrmacht bei. *PRESSEMITTEILUNG, Die LINKE im Bundestag* [Online], 15 May 2009. Available at: http://www.linksfraktion.de/pressemitteilungen/bundeswehr-traegt-aktiv-verklaerung-wehrmacht/.

Jensby, S.A. 2002. Brev til de Blå Baretter, Landsformand Erik Blyme. *De Blå Baretter* [Online], 2 November 2002. Available at: http://www.blaabaretter.dk/files/manager/monument/forsvarsministerens%20svar%20p%C3%A3%C2%A5%20indstilling%20%20nov%202002.pdf [accessed: 15 November 2011].

Jensen, C. 2010. Hvorfor er vi tavse om krigen i Afghanistan? *Politiken* [Online], 22 May 2010. Available at: http://politiken.dk/debat/ECE977458/hvorfor-er-vi-tavse-om-krigen-i-afghanistan/ [accessed: 23 May 2010].

Jensen, C. 2013. Hurra – vi tabte! *Politiken* [Online], 2 January 2013. Available at: http://politiken.dk/debat/kroniker/ECE1856213/hurra---vi-tabte/ [accessed: 2 January 2013].

Jensen, T. 2012. Krig og kruspersille. *Berlingske* [Online], 5 September 2012. Available at: http://tomjensen.blogs.berlingske.dk/2012/09/05/krig-og-kruspersille/ [accessed: 7 November 2012].

Jentleson, B.W. and Britton, R.L. 1998. Still Pretty Prudent: Post-Cold War American Public Opinion on the Use of Military Force. *Journal of Conflict Resolution*, 42(4): 395–417.

Johannesen, L.U. 2008. *De danske tigre. Med livet som indsats i Afghanistan.* København: Gyldendal.

Johnson, D.D.P. and Tierney, D. 2006. *Failing to Win: Perceptions of Victory and Defeat in International Politics.* Cambridge: Harvard University Press.

Jung, F.J. 2007. Ein Ehrenmal ist überfällig. *Die Welt* [Online], 23 June 2007. Available at: http://www.welt.de/debatte/kommentare/article6069245/Ein-Ehrenmal-ist-ueberfaellig.html [accessed: 13 March 2013].

Jung, F.J. 2009. Im Schmerz nicht allein. 2 July 2009, Berlin: Bundesministerium der Verteidigung.

Junge, B. 2009. Kein Krieg, also auch kein Ehrengrab. *Der Tagesspiegel* [Online], 8 September 2009. Available at: http://www.tagesspiegel.de/politik/international/afghanistan-keinkrieg-also-auch-keinehrengrab/1595764.html [accessed: 19 May 2010].

Jungholt, T. 2009. Das ist auch die Angst vor der Geschichte. *Die Welt* [Online], 7 September 2009. Available at: http://www.welt.de/welt_print/politik/article4477247/Das-ist-auch-die-Angst-vor-der-Geschichte.html [accessed: 14 March 2013].

Jungholt, T. 2011. Trauer und Misstrauen. *Die Welt* [Online], 26 February 2011. Available at: http://www.welt.de/print/die_welt/debatte/article12647976/Trauer-und-Misstrauen.html [accessed: 12 March 2012].

Jyllands-Posten. 2001a. Al Qaeda knust i Tora Bora [Online], 16 December 2001. Available at: http://jp.dk/udland/article620435.ece [accessed: 22 September 2011].

Jyllands-Posten. 2001b. Fogh Rasmussen udsteder blankocheck i terrorkrig [Online], 17 December 2001. Available at: http://jp.dk/udland/article620615.ece [accessed: 9 June 2011].

Jyllands-Posten. 2001c. Forsker: Militært tilbud til USA er gratis for Fogh [Online], 6 December 2001. Available at: http://jp.dk/udland/article626449.ece [accessed: 22 November 2011].

Jyllands-Posten. 2007. Kamp for flagdag [Online], 29 August 2007. Available at: http://jyllands-posten.dk/indland/article3461490.ece [accessed: 17 March 2011].

Jyllands-Posten. 2009. Per Stig: I hjælper Taleban [Online], 4 November 2009. Available at: http://jyllands-posten.dk/politik/article4262670.ece [accessed: 16 December 2012].

Jyllands-Posten. 2010a. Danskerne er blevet krigstrætte [Online], 24 July 2010. Available at: http://politiken.dk/indland/ECE1023029/danskerne-er-blevet-krigstraette/ [accessed: 22 August 2011].

Jyllands-Posten. 2010b. Fogh advarer mot dalende Afghanistanstøtte [Online], 13 July 2010. Available at: http://jp.dk/udland/article2124004.ece [accessed: 19 September 2011].

Jyllands-Posten. 2011. Drab på danskere forsvares [Online], 21 January. Available at: http://jyllands-posten.dk/indland/article4519951.ece [accessed: 10 March 2012].

Jyllands-Posten. 2012. Afghanistan [Online], 15 March 2012. Available at: http://jyllands-posten.dk/opinion/leder/article4651113.ece [accessed: 14 March 2013].

Jörges, H.-U. 2007. Unsere versteckten Toten. *Stern* [Online], 11 January 2007. Available at: http://www.stern.de/politik/deutschland/zwischenruf/zwischenruf-unsere-versteckten-toten-579918.html [accessed: 19 September 2011].

Kaim, M. 2009. Deutsches Interesse versus Bündnisverpflichtung: Zur Frage nationaler Handlunsgspielräume bei Auslandseinsätzen der Bundeswehr. In Gießmann, H.J. and Wagner, A. (eds) *Armee im Einsatz*. Baden-Baden: Nomos.

Kaiser, A. 2010. *Von Helden und Opfern. Eine Geschichte des Volkstrauertags*. Frankfurt: Campus Verlag.

Kant, I. 2003. *To Perpetual Peace. A Philosophical Sketch*. Indianapolis: Hackett.

Katzenstein, P.J. 1998. *Cultural Norms and National Behaviour: Police and Military in Postwar Japan*. Ithaca: Cornell University Press.

Keil, L.-B. 2010. Bürger vertrauen Militär mehr als der Politik. *Die Welt* [Online], 18 February 2010. Available at: http://www.welt.de/welt_print/politik/article6445347/Buerger-vertrauen-Militaer-mehr-als-der-Politik.html.

Kemp, R. and Hughes, R. 2009. *Attack State Red*. London: Penguin.

Kennedy, M. 2009. Remembrance Sunday: Rising numbers gather at Cenotaph to pay tribute. *The Guardian* [Online], 8 November 2009. Available at: http://www.guardian.co.uk/uk/2009/nov/08/remembrance-sunday-cenotaph [accessed: 18 June 2011].

Keren, M. 2009. Introduction. In Keren, M. and Herwig, H.H. (eds) *War Memory and Popular Culture*. Jefferson: McFarland & Company, 1–7.

Key, V.O. 1961. *Public Opinion and American Democracy.* New York: Alfred A. Knopf.

Kienholtz, E. 1996. The Portable War Memorial. In Stiles, K. and Selz, P.H. (eds) *Theories and Documents of Contemporary Art: A Sourcebook of Artists' Writings.* Berkely: University of California Press, 514.

Kiley, S. 2009. *Desperate Glory. At War in Helmand with Britain's 16 Air Assault Brigade.* London: Bloomsbury.

King, A. 1998. *Memorials Of The Great War. The Symbolism and Politics of Remembrance.* Oxford: Berg.

King, A. 2010. The Afghan War and 'postmodern' memory: Commemoration and the dead of Helmand. *The British Journal of Sociology*, 61(1): 1–25.

King, L. 2009. New U.S. tactics said to reduce Afghan civilian deaths. *Los Angeles Times* [Online], 28 August 2009. Available at: http://articles.latimes.com/2009/aug/28/world/fg-afghan-civilians28 [accessed: 20 January 2012].

Klarskov, K. and Svane, A.M. 2007. Fogh: Villy Søvndal driver plat på soldats død. *Jyllands-Posten* [Online], 15 May 2007. Available at: http://www.jyllands-posten.dk/protected/premium/indland/article3878530.ece.

Klein, S. 2008. Vielleicht waren wir ein bisschen naiv. *Süddeutsche Zeitung*, 15 October 2008, 1.

Knudsen, A. 2011. Nutiden genopført. *Weekendavisen.*

Koelbl, H. 2011. Was ist los an der Front? *Zeit Magazin*, 1 December 2011, 11–46.

Koelbl, S. and Szandar, A. 2008. Der wichtigste Feind. *Spiegel* [Online], 19 May 2008. Available at: http://www.spiegel.de/spiegel/print/d-57038064.html [accessed: 11 March 2013].

Koldehoff, S. 2009. Es ist 'ein Vermeidungsdenkmal'. dradio.de [Online], 8 September 2009. Available at: http://www.dradio.de/dlf/sendungen/kulturheute/1031189/drucken/ [accessed: 19 November 2011].

Kornelius, S. 2009. *Der unerklärte krieg. Deutschlands Selbstbetrug in Afghanistan.* Hamburg: Edition Körber-Stiftung.

Koselleck, R. 1979. Kriegerdenkmäle als Identitätsstiftungen der Überlebenden. In Marquard, O. and Stierle, K.-H. (eds) *Identität.* München: Fink Verlag, 255–75.

Koselleck, R. 1994. Einleitung. In Koselleck, R. and Jeismann, M. (eds) *Der Politische Totenkult. Kriegerdenkmäler in der Moderne.* München: Wilhelm Fink Verlag, 9–20.

Koselleck, R. and White, H. 2002. *The Practice of Conceptual History: Timing History, Spacing Concepts.* Stanford, CA: Stanford University Press.

Koureas, G. 2007. *Memory, Masculinity and National Identity in British Visual Culture 1914–1930.* Aldershot: Ashgate.

Kremp, H. 2005. Tod in Afghanistan. *Welt*, 26 November 2005.

Kriner, D. and Wilson, G. 2010. *Elites, Events and British Support for the War in Afghanistan.* Washington DC: Paper prepared for presentation at the annual meeting of the American Political Science Association.

Kristeligt Dagblad. 2011. Danskerne: Afghanistan ikke prisen værd. *Kristeligt Dagblad* [Online], 11 October 2011. Available at: http://www.kristeligt-dagblad.dk/artikel/436191:Danmark--Danskerne--Afghanistan-ikke-prisen-vaerd [accessed: 12 October 2011].

Kristensen, K. 2007. Elsket af sit kompagni. *Hæren. Den Kongelige Livgarde* [Online], 16 October 2007. Available at: http://forsvaret.dk/LG/Nyt%20og%20Presse/Nyheder/Pages/Elsket%20af%20sit%20kompagni.aspx [accessed: 19 November 2012].

Krog, T.N. 2010. Dansk soldat: Selvfølgelig jubler vi efter et drab. BT [Online], 18 May 2010. Available at: http://www.bt.dk/danmark/dansk-soldat-selvfoelgelig-jubler-vi-efter-et-drab [accessed: 20 November 2010].

Kull, S. 1995. Misreading the Public Mood. *Bulletin of the Atomic Scientists*, 51(2): 55–9.

Kull, S. and Destler, I.M. 1999. *Misreading the Public: The Myth of a New Isolationism.* Washington DC: Brookings.

Kull, S., Ramsay, C. and Lewis, E. 2003–2004. Misperceptions, the Media, and the Iraq War. *Political Science Quarterly*, 118(4): 569–98.

Kurtz, M.J. 2006. *America and the Return of Nazi Contraband: The Recovery of Europe's Cultural Treasures.* Cambridge: Cambridge University Press.

Kümmel, G. and Leonhard, N. 2004. Casualty Shyness and Democracy in Germany. *Sicherheit und Frieden*, 22(3): 119–26.

Köhler, H. 2005. Rede von Bundespräsident Horst Köhler auf der Kommandeurtagung der Bundeswehr am 10. Oktober 2005 in Bonn. *Bulletin* [Online], 10 November 2005, Nr. 79-1. Available at: http://www.bundesregierung.de [accessed: 23 May 2011].

König, J. 2010. Recht und Einsatz. 'Von Rechtsunsicherheit kann keine Rede sein'. Interview mit Dr. Dieter Weingärtner, Abteilungsleiter Recht im Bundesministerium der Verteidigung. *Kompas*, 3: 9–10.

Lahrmann, H. 2009. Opbakning til dansk indsats i Helmand. *Hovedorganisationen af Officerer i Danmark* [Online], 4 November 2009. Available at: http://www.hod.dk/index.dsp?page=1138 [accessed: 13 December 2012].

Laïdi, Z. 2008. *La norme sans la force.* Paris: SciencesPo. Les Presses.

Laing, C. 2012. 'Green on blue' in Helmand – what the international media don't tell you. *UK Forces Afghanistan* [Online], 29 May 2012. Available at: http://ukforcesafghanistan.wordpress.com/2012/05/29/green-on-blue-in-helmand-what-the-international-media-dont-tell-you/ [accessed: 1 June 2012].

Lamerton, R. 1973. *Care of the Dying.* London: Priory Press.

Larsen, R.E. 2013. Død for Danmark – i Afghanistan? *Politiken* [Online], 4 January 2013. Available at: http://politiken.dk/debat/profiler/engelbreth/ECE1858468/doed-for-danmark---i-afghanistan/ [accessed: 4 January 2013].

Larsen, T. 2008. *Dagbog fra Afghanistan.* København: Borgen.

Larson, E.V. 1996. *Casualties and Consensus. The Historical Role of Casualties in Domestic Support for U.S. Military Operations.* Santa Monica: RAND.

Lassen, M.K. 2009. *Soldat i Afghanistan. En hverdag ved frontlinjen.* Hillerød: Fokus forlag.

Lavery, C. 2001. DIVIDED Scots split on bombing; UNITED Bin Laden must face trial. *Sunday Mail*, 21 October.

Lavrsen, L. 2010. Danmark er en krigsnation. *Information* [Online], 18 June 2010. Available at: http://www.information.dk/236879 [accessed: 22 June 2010].

Lawrence, R. 2000. Game-Framing the Issues: Tracking the Strategy Frame in Public Policy News. *Political Communication*, 17(2): 93–114.

Le Monde. 2001. Nous sommes tous Américains. *Le Monde*, 13 September.

Leersch, H.-J. 2002. Berlin ist empört über amerikanische Offenheit. *Welt* [Online], 5 March 2002. Available at: http://www.welt.de/print-welt/article377563/Berlin-ist-empoert-ueber-amerikanische-Offenheit.html.

Leersch, H.-J. 2005. Vergessene Soldaten. *Welt*, 29 September 2005.

Leersch, H.-J. 2007. Man kann auch falsch aufklären. *Welt*, 9 February 2007.

Lindner, A.M. 2008. Controlling the Media in Iraq. *Contexts*, 7(2): 32–8.

Lindquist, A. 2011. Højrefløj og medier kom tidligt til Hizb-ut-Tahrir-moddemonstration. *Politiken* [Online], 21 January. Available at: http://politiken.dk/indland/ECE1174025/hoejrefloej-og-medier-kom-tidligt-til-hizb-ut-tahrir-moddemonstration/ [accessed: 8 March 2011].

Livgarden. 2008. Kompagniet's gave til Anders Storruds familie. Onsdag d. 2 juli tog dele af 1 kompagni på havearbejde. Garderportal.dk [Online], 7 July 2008. Available at: http://www.garderportal.dk/AndersStorrudGaveTilFamilienFraFoersteKompagni.html [accessed: 10 May 2010].

Livingston, S. 1997. *Clarifying the CNN Effect: An Examination of Media Effects According to Type of Military Intervention.* Harvard University: John F. Kennedy School of Government's Joan Shorenstein Center on the Press, Politics and Public Policy.

Llewellyn, K.R. 2012. *Remembering The Dead? What Do War Memorials in Essex Say About The Great War*. University of Essex.

Lloyd, D. 1998. *Battlefield Tourism. Pilgrimage and Commemoration of the Great War in Britain, Canada and Australia.* Oxford: Berg.

Lohse, G. 2011. Istedløven vender hjem til Flensborg. *DR Kultur* [Online], 9 September 2011. Available at: http://www.dr.dk/Nyheder/Kultur/2011/09/09/09094117.htm [accessed: 12 September 2011].

Luhmann, N. 2004. *Die Realität der Massenmedien.* Opladen: VS Verlag für Sozialwissenschaften.

Lund, K. 2013. Regeringen: Krigen i Afghanistan var umulig at vinde. *Politiken* [Online], 29 January 2013. Available at: http://politiken.dk/politik/ECE1882421/regeringen-krigen-i-afghanistan-var-umulig-at-vinde/ [accessed: 30 January 2013].

Luttwak, E.N. 1994. Where Are the Great Powers? At Home with the Kids? *Foreign Affairs*, 73(4): 23–8.

Luttwak, E.N. 1996. A Post-Heroic Military Policy. *World Affairs*, 75(4): 33–44.

Lüders, S. 2006. Liegt Deutschlands Zukunft in den Händen der Bundeswehr? *Humanistische Union – Pressemitteilung*, 26 October.

Löwenstein, S. 2008. Helden im Geplankel. *Frankfurter Allgemeine Zeitung* [Online], 16 October 2008. Available at: http://www.faz.net/aktuell/politik/inland/bundestag-und-deutsche-soldaten-helden-im-geplaenkel-1707707.html.

Macintyre, J. 2008. Soldiers' deaths 'down to lack of basic equipment'. *The Independent*, 16 February.

Mack, A.J.R. 2005. Peace on Earth? Increasingly Yes! *Washington Post* [Online], 28 December 2005. Available at: http://www.washingtonpost.com/wp-dyn/content/article/2005/12/27/AR2005122700732.html [accessed: 16 June 2011].

Mannitz, S. 2011. Redefining Soldierly Role Models in Germany. *Armed Forces & Society*, 37(4): 680–700.

Mans, U., Meindersma, C. and Burema, L. 2008. *Eyes Wide Shut? The Impact of Embedded Journalism on Dutch Newspaper Coverage of Afghanistan.* The Hague: The Hague Centre for Strategic Studies.

Martinsen, K.D. 2010. National interests in German security white books. *National Identities*, 12(2): 161–80.

Mason, T. and Ashton, E. 2009. Troops have enough to do job, says Brown. *The Independent*, 23 July.

Matern, T. 2012. Nato-Truppenabzug aus Afghanistan: Warum Karsai zu kühn ist. *Süddeutsche Zeitung* [Online], 15 March 2012. Available at: http://www.sueddeutsche.de/politik/nato-truppenabzug-aus-afghanistan-warum-karsai-zu-kuehn-ist-1.1310281 [accessed: 16 March 2012].

Matlary, J.H. 2009. *European Union Security Dynamics: In the New National Interest.* Basingstoke: Palgrave Macmillan.

Maull, H.W. 2007. Deutschland als Zivilmacht. In Schmidt, S., Hellmann, G. and Wolf, R. (eds) *Handbuch zur deutschen Außenpolitik.* Wiesbaden: VS Verlag für Sozialwissenschaften, 73–84.

Mayntz, G. 2008. Was bedeutet Parlamentsarmee? *Bundestag und Bundeswehr*, 1–5.

McCormick, C. 1975. The Undiluted Wisdom of Marshall McLuhan. *Montreal Gazette*, 16 May 1975.

McIntyre, C. 1994. *Monuments of War.* London: Robert Hale.

McNab, A. (ed.) 2009. *SPOKEN FROM THE FRONT. Real Voices from the Battlefields of Afghanistan.* London: Corgi Books.

Mead, W.R. 2004. *Kisses and Kicks: German-American Relations in the Age of Bush.* Washington DC: AICGS.

Meltzer, T. 2009. Can't Gordon Brown do anything right? *The Guardian* [Online], 10 November 2009. Available at: http://www.guardian.co.uk/politics/2009/nov/10/gordon-brown-do-anything-right/print [accessed: 28 April 2011].

Merchet, J.-D. 2009. *Défense européenne, la grande illusion.* Paris: Larousse.

Merkel, A. 2009. Rede von Bundeskanzlerin Dr. Angela Merkel beim feierlichen Gelöbnis der Bundeswehr vor dem Reichstagsgebäude am 20. Juli 2009 in Berlin. Berlin: Bundesregierung.

Merkel, A. 2010a. Regierungserklärung von Bundeskanzlerin Dr. Angela Merkel zum Einsatz der Bundeswehr in Afghanistan, 22 April 2010. *BULLETIN DER BUNDESREGIERUNG*, 42–1.

Merkel, A. 2010b. Regierungserklärung von Bundeskanzlerin Merkel zum Einsatz der Bundeswehr in Afghanistan, 22 April 2010. Available at: http://www.bundesregierung.de/Content/DE/Regierungserklaerung/2010/2010-04-22-merkel-erklaerung-afghanistan.html [accessed: 23 April 2010].

Merkel, A. 2010c. Soldaten Rückhalt geben. Regierungserklärung von Bundeskanzlerin Dr. Angela Merkel (CDU)(CSU) 22 April 2010. Berlin: Deutscher Bundestag.

Merkur. 2010. Rufe an Teilnahme Merkels an Trauerfeier [Online], 8 April 2010. Available at: http://www.merkur-online.de/nachrichten/politik/rufe-nach-teilnahme-merkels-trauerfeier-706079.html [accessed: 9 May 2011].

Merrin, W. 2005. *Baudrillard and the Media. A Critical Introduction*. Cambridge: Polity Press.

Merson, J. 2008. *War Lessons: How I Fought To Be a Hero and Learned That War Is Terror*. Enfield: Publishers group.

Meyer, B. 2009. *Innere Führung und Auslandseinsätze: Was wird aus dem Markenzeichen der Bundeswehr*. Frankfurt am Main: Hessische Stiftung Friedens- und Konfliktforschung.

Meyer, B. 2010. *Bundeswehr ohne Wehrpflichtige. Was folgt daraus für die Parlamentsarmee im Einsatz?* Franfurt: Hessische Stiftung Friedens und Konfliktforschung.

Meyer, C.O. 2006. *The Quest for a European Strategic Culture: Changing Norms on Security and Defence in the European Union*. Hampshire: Palgrave Macmillan.

Meyer, S. 2011a. Meine Soldaten sind meine Helden des Alltags. *Die Welt* [Online], 6 September 2011. Available at: http://www.welt.de/politik/deutschland/article13586525/Meine-Soldaten-sind-meine-Helden-des-Alltags.html [accessed: 10 March 2013].

Meyer, S. 2011b. Sie fehlen unserem Land. *Die Welt* [Online], 4 June 2011. Available at: http://www.welt.de/print/die_welt/politik/article13411590/Sie-fehlen-unserem-Land.html [accessed: 5 June 2011].

Meyer, S. 2011c. 'Wir gehen dorthin, wo es Menschen etwas bringt'. *Die Welt* [Online], 1 April 2011 [accessed: 2 April 2011].

Meyer, S. 2012. ISAF-General: Übergabe in Afghanistan ist zu schaffen. *Bundeswehr* [Online], 30 May 2012. Available at: http://www.bundeswehr.de [accessed: 1 June 2012].

Michalski, M. and Gow, J. 2007. *War, Image and Legitimacy: Viewing Contemporary Conflict*. London: Routledge.

Michelis, H. 2010. KSK-Kommandeur Ammon im Gespräch: Töten gehört zum Auftrag. *Rheinische Post* [Online], 20 May 2010. Available at: http://www.rp-online.de/politik/deutschland/toeten-gehoert-zum-auftrag-1.2294527 [accessed: 16 October 2012].

Mindet. 2007. Til minde om Anders Johan Stæhr Storrud, major. 2007. Available at: http://mindet.dk/minde/anders-johan-staehr-storrud-major [accessed: 10 March 2013].

Minkmar, N. 2008. Krieg ohne Opfer. *Frankfurter Allgemeine Zeitung* [Online], 16 November 2008. Available at: http://www.faz.net/aktuell/feuilleton/debatten/zum-volkstrauertag-krieg-ohne-opfer-1726353.html [accessed: 12 March 2013].

Mittig, H.-E. 2007. Blech und Stein. *Frankfurter Rundschau* [Online], 26 October 2007. Available at: http://www.fr-online.de/doku---debatte/blech-und-stein,1472608,2852032.html [accessed: 13 March 2013].

Mittig, H.-E. 2008. Sind demokratische Werte monumental repräsentierbar? In Hettling, M. and Echternkamp, J. (eds) *Bedingt erinnerungsbereit. Soldatengedenken in der Bundesrepublik.* Göttingen: Vandenhoeck & Ruprecht, 132–48.

MoD. 2009. First Elizabeth Cross presented to family of fallen soldier. *COMMANDO Veterans Association* [Online], 18 August 2009. Available at: http://www.commandoveterans.org/cdoForum/posts/list/1041.page [accessed: 11 March 2012].

MoD. 2011a. Dr Fox – we must continue to battle Al-Qaeda. *Defence News* [Online], 3 May 2011. Available at: http://www.mod.uk/DefenceInternet/DefenceNews/DefencePolicyAndBusiness/DrFoxWeMustContinueToBattleAlqaeda.htm [accessed: 6 November 2011].

MoD. 2011b. Lance Corporal Liam Richard Tasker killed in Afghanistan. 2 March 2011. Available at: http://www.mod.uk/DefenceInternet/DefenceNews/MilitaryOperations/LanceCorporalLiamRichardTaskerKilledInAfghanistan.htm [accessed: 3 March 2011].

MoD. 2013a. British fatalities. Operations in Afghanistan. London.

MoD. 2013b. Op Herrick casualty and fatality tables [Online], 14 February 2013. Available at: https://www.gov.uk/government/uploads/system/uploads/attachment_data/file/82766/20130213_Op_Herrick_to_31_Jan_13.pdf [accessed: 15 March 2015].

Mogg, U. and Bartels, H.-P. 2007. Die 'gute' und die 'böse' Mission. Wie weiter mit der Operation 'Enduring Freedom'? *Die Welt* [Online], 17 November 2007. Available at: http://www.hans-peter-bartels.de/pdf/241.pdf [accessed: 13 March 2013].

Mollerup, J. 2007. Krigen om troværdigheden. *DR* [Online], 26 January 2007. Available at: http://www.dr.dk/OmDR/Lytternes_og_seernes_redaktoer/Klummer/2007/0126150316.htm [accessed: 18 May 2011].

Morgan, L.R. 2010. January 2007. In Mcnab, A. (ed.) *SPOKEN FROM THE FRONT. Real Voices from the Battlefields of Afghanistan.* London: Corgi Books, 184–7.

Morris, S. 2009. Prince of Wales and prime minister lead tributes to Lt Col Rupert Thorneloe. *The Guardian* [Online], 3 July 2009. Available at: http://www.guardian.co.uk/world/2009/jul/03/tributes-officer-killed-afghanistan-thorneloe?intcmp=239 [accessed: 9 March 2011].

Mosse, G.L. 1990. *Fallen Soldiers. Reshaping the Memory of the World Wars.* New York: Oxford University Press.

Mueller, J. 1971. Trends in Popular Support for the Wars in Korea and Vietnam. *American Political Science Review*, 65(1): 358–75.

Mueller, J. 1994. *Policy and Opinion in the Gulf War.* Chicago: The University of Chicago Press.

Mueller, J. 2004. *The Remnants of War.* Ithaca: Cornell University Press.

Mueller, J. 2011. *War and Ideas. Selected Essays.* London: Routledge.

Mueller, J.E. 1973. *War, Presidents, and Public Opinion.* New York: Wiley.

Müller, A. 2012. Todesfälle im Auslandseinsatz [Online], 18 January 2010. Available at: http://www.bundeswehr.de/portal/ [accessed: 20 January 2010].

Müller, H., Fey, M., Mannitz, S. and Schörning, N. 2010. *Demokratie, Streitkräfte und militärische Einsätze: Der 'zweite Gesellschaftsvertrag' steht auf dem Spiel.* Frankfurt: Hessische Stiftung Friedens – und Konfliktforschung.

Müller, P. 2006. Deutschlands vergessene Soldaten – warum ihr Krieg auch in der Heimat immer weitergeht. *Welt am Sonntag*, 17 December.

Münch, P. 2009. Freund oder Feind? Zur Einschätzung von Sicherheitsbedrohungen in Friedensmissionen am Beispiel der Bundeswehr in Afghanistan. *Arbeitspapier.* Institut für Politikwissenschaft, Universität Hamburg.

Münkler, H. 2007. Opfer und Erinnerung. *Der Tagesspiegel* [Online], 24 November 2007. Available at: http://www.tagesspiegel.de/kultur/opfer-und-erinnerung-fuer-die-gute-sache/1103658.html [accessed: 16 April 2012].

Møller, B. 2000. Dansk udenrigspolitik efter Kosovo. *Politica*, 32(2): 172–85.

Møller, K.M. 2009. *Stille bøn i Helmand. Om danske soldaters erfaring, samvittighed og tro.* Store Heddinge: ZAR.

Møller, L.R. 2010. *Vi slår ihjel og lever med det.* København: Informations forlag.

Møller, P.S. 2006. Forslag til folketingsbeslutning om udvidelse af det danske bidrag til den internationale sikkerhedsstyrke ISAF i Afghanistan, 2 February 2006. København: Folketinget.

n-tv. 2010. Polizeiausbildung in Afghanistan. De Maizière räumt Fehler ein [Online], 28 March 2010. Available at: http://www.n-tv.de/politik/De-Maizi-re-raeumt-Fehler-ein-article798108.html [accessed: 22 May 2011].

Nachtwei, W. 2007. Rede Nachtweis zur Bundeswehr-Ehrenmal-Debatte. 6 July 2007. Available at: http://www.nachtwei.de/index.php/articles/news/546 [accessed: 17 November 2011].

Nagle, J. 2012. Not losing in Afghanistan. *Washington Post* [Online], 19 October 2012. Available at: http://articles.washingtonpost.com/2012-10-19/opinions/35500279_1_security-forces-tribal-regions-islamabad [accessed: 20 October 2012].

National Commission 2004. The 9/11 Commission Report. Washington DC: US Government.

Naumann, K. 2005. Abwehr, Abschreckung, Distanzierung, Militär, Öffentlichkeit und Tod in der Bundesrepublik. In Hettling, M. and Echterkamp, J. (eds)

Bedingt erinnerungsbereit. Soldatengedenken in der Bundesrepublik. Göttingen: Vandehoeck & Ruprecht, 162–74.

Neues Deutschland. 2010. Trauerfeier für Bundeswehr-Soldaten. 10 April 2010, 4.

Newman, B.I. and Perloff, R.M. 2004. Political Marketing: Theory, Research and Application. In Kaid, L.L. (ed.) *Handbook of Political Communication Research.* Mahwah, N.J.: Lawrence Erlbaum Associates, 17–43.

Nicol, M. 2013. Top commander: If UK leaves Afghanistan, Taliban will return to launch global jihad. *Daily Mail* [Online], 2 February 2013. Available at: http://www.dailymail.co.uk/news/article-2272585/Top-commander-If-UK-leaves-Afghanistan-Taliban-return-launch-global-jihad.html#axzz2KIyxaVrO [accessed: 5 February 2013].

Nielsen, A. 2006. Forsvaret kan ikke bekræfte danske civildrab. *Politiken* [Online], 21 September 2006. Available at: http://politiken.dk/udland/article176434.ece [accessed: 17 November 2011].

Nielsen, C.N. 2012. Har vores indsats i Irak og Afghanistan været forgæves? *Politiken* [Online], 22 December 2012. Available at: http://politiken.dk/debat/ECE1851048/har-vores-indsats-i-irak-og-afghanistan-vaeret-forgaeves/ [accessed: 12 January 2013].

Nielsen, J. 2008. Hyldest til soldater *Hæren. Garderhusarregimentet* [Online], 8 May 2008. Available at: http://forsvaret.dk/GHR/Nyt%20og%20Presse/Oevrige%20nyheder/Pages/Hyldesttilsoldater.aspx [accessed: 17 March 2011].

Noelle-Neumann, E. 1974. The Spiral of Silence: A Theory of Public Opinion. *Journal of Communication*, 24(2): 43–51.

Nolte, P. 2008. Fremde Soldaten. Deutschlands Nichtverhältnis zu seiner Armee. *Der Spiegel*, 48.

Nora, P. 1989. Between Memory and History: Les Lieux de Mémoire. *Representations*, Spring, 26: 3–24.

Norris, P., Kern, M. and Just, M. 2003. Framing Terrorism. In Norris, P., Kern, M. and Just, M. (eds) *Framing Terrorism. The News Media, the Government, and the Public.* London: Routledge, 3–23.

Norton-Taylor, R. 2009. Ministers indecisive as public opinion shifts after deaths of soldiers in Afghanistan. *The Guardian*, 10 July.

Norton-Taylor, R., Glover, J. and Watt, N. 2009. Public support for war in Afghanistan is firm, despite deaths. *The Guardian* [Online], 13 July 2009. Available at: http://www.guardian.co.uk/uk/2009/jul/13/afghanistan-war-poll-public-support [accessed: 15 May 2012].

Olesen, G. 2011. Anmeldelse: 'Den fjerne krig – dansk soldat i Afghanistan'. *RIKO – Rådet for international konfliktløsning* [Online], 9 October 2011. Available at: http://riko.nu/2011/10/anmeldelse-%E2%80%9Dden-fjerne-krig-%E2%80%93-dansk-soldat-i-afghanistan%E2%80%9D/ [accessed: 11 October 2011].

Overhaus, M. 2009. *Die deutsche NATO-Politik. Vom Ende des Kalten Krieges bis zum Kampf gegen den Terrorismus.* Baden-Baden: Nomos Verlag.

Oxfordshire County Council. 2012. Repatriation of fallen service personnel. 3 September 2012. Available at: http://www.oxfordshire.gov.uk/cms/content/repatriation-fallen-service-personnel [accessed: 10 March 2013]

Pau, P. 2003. Am Hindukusch gilt kein Grundgesetz. *Die Linke Landesverband Berlin* [Online], 21 May. Available at: http://www.die-linke-berlin.de/nc/politik/presse/detail/artikel/am-hindukusch-gilt-kein-grundgesetz/ [accessed: 20 December 2012].

Pedersen, J.S. 2010. Søren Gade kom forbi. *Hæren. Jyske Dragonregiment* [Online], 6 October 2010. Available at: http://forsvaret.dk/JDR/Nyt%20og%20Presse/Nyheder%20fra%20JDR/Pages/Netv%C3%A6rkscafed2september2010.aspx [accessed: 10 October 2010].

Petersen, R. 2012. *Den bedste ambassadør. Civil-militære relationer og demokratisk kontrol i Danmark 1991–2011*. Roskilde: RUC.

Peterson, F. n.d. Uhland-Gedichte. *German Language* [Online]. Available at: http://german.about.com/library/blkameraden.htm [accessed: 13 March 2013].

Pike, H. 2008. *From the Front Line. Family Letters and Diaries – 1900 to the Falklands and Afghanistan*. Barnsley: Pen & Sword.

Pinker, S. 2007. A History of Violence. We're getting nicer every day. *The New Republic*, March, 18–21.

Pinker, S. 2011. *The Better Angels of Our Nature*. London: Allen Lane.

Politiken. 2006. Danskere har dræbt mindst 25 talebanere. *Politiken* [Online], 29 August 2006. Available at: http://politiken.dk/udland/ECE166757/danskere-har-draebt-mindst-25-talebanere/ [accessed: 13 March 2013].

Politiken. 2011a. Danske soldater døde i Afghanistan. *Politiken* [Online], 4 January 2011. Available at: http://politiken.dk/udland/internationalt620px/ECE1167315/danske-soldater-doede-i-afghanistan/ [accessed: 15 February 2011].

Politiken. 2011b. Danske soldater vil blive i Afghanistan. *Politiken* [Online], 5 January 2011. Available at: http://politiken.dk/indland/ECE1159678/danske-soldater-vil-blive-i-afghanistan/ [accessed: 6 January 2011].

Politiken. 2011c. Soldater raser over omstridt Hizb-ut-Tahrir-møde. *Politiken* [Online], 18 January 2011. Available at: http://politiken.dk/indland/ECE1170789/soldater-raser-over-omstridt-hizb-ut-tahrir-moede/ [accessed: 10 March 2013].

Politiken. 2009. Soldater: Afghanistan-mission er som en hullet robåd. *Politiken* [Online], 4 November 2009. Available at: http://politiken.dk/indland/ECE825588/soldater-afghanistan-mission-er-som-en-hullet-robaad/ [accessed: 16 December 2012].

Posen, B. 2000. The War for Kosovo: Serbia's Political Military Strategy. *International Security*, 24(4): 39–84.

Press Association. 2012. 73% say Afghan war is unwinnable [Online], 20 April 2012. Available at: http://www.pressassociation.com/component/pafeeds/2012/03/12/73_say_afghan_war_is_unwinnable?camefrom=home [accessed: 17 March 2012].

Pries, K. and Meng, R. 2006. Unser Beitrag wird respektiert. *Frankfurter Rundschau*, 28 November 2006, 1–2.

Prince, R. 2009. Flag launched to mark Armed Forces Day. *The Telegraph* [Online], 28 April 2009. Available at: http://www.telegraph.co.uk/news/uknews/defence/5230627/Flag-launched-to-mark-Armed-Forces-Day.html [accessed: 1 November 2011].

Quist, N.O. 2008. Christian fik en militær begravelse. BT [Online], 11 April 2008. Available at: http://www.bt.dk/nyheder/christian-fik-en-militaer-begravelse.

RAF. 2011. Details of repatriations through RAF Brize Norton and Carterton. *RAF Police Association* [Online], 8 September 2011. Available at: http://www.rafpa.com/carterton.htm [accessed: 9 June 2012].

Ramussen, A.F. 2003. Hvad kan det nytte? *Berlingske Tidende* [Online], 26 March 2003. Available at: http://www.stm.dk/_p_5364.html [accessed: 22 May 2010].

Ramussen, P.E. 2006. *Soldat for Danmark. Oversergent Jakob Birger Hansen.* København: Aschehoug.

Rasmussen, A.F. 2003. Tale holdt i forbindelse med fejringen af 60-året for ophøret af samarbejdspolitikken [Online], 29 August 2003. Available at: http://danmarkshistorien.dk/leksikon-og-kilder/vis/materiale/anders-fogh-rasmussen-v-om-samarbejdspolitikken-29-august-2003/ [accessed: 22 May 2010].

Rasmussen, A.F. 2006a. Fogh: Danmark må gøre op med småstatsmentaliteten. *Mandag Morgen* [Online], 11 September 2006 [accessed: 22 May 2012].

Rasmussen, A.F. 2006b. Statsminister Anders Fogh Rasmussens festforelæsning ved Forsvarsakademiets årsdag den 1. November 2006. 1 November 2006. Available at: http://www.stm.dk/_p_7537.html [accessed: 20 May 2010].

Rasmussen, A.F. 2007a. Statsminister Anders Fogh Rasmussens nytårstale 2007. *Statsministeriet* [Online], 1 January 2007. Available at: http://www.stm.dk/_p_7551.html [accessed: 6 March 2010].

Rasmussen, A.F. 2007b. Statsminister Anders Fogh Rasmussens tale ved Folketingets afslutningsdebat 31. maj 2007. *Statsministeriet* [Online], 31 May 2007. Available at: http://www.stm.dk/_p_7562.html [accessed: 9 March 2011].

Rasmussen, L.L. 2009. Statsministerens tale ved mindegudstjeneste i Holmens Kirke for Danmarks faldne, den 5. September 2009. 5 September 2009. Available at: http://www.stm.dk/_p_12921.html [accessed: 9 May 2011].

Rasmussen, L.L. 2010. Statsministerens tale ved mindegudstjenesten i Holmens kirke for Danmarks faldne den 5. September 2010. *Statsminsteriet* [Online], 5 September 2010. Available at: http://www.stm.dk/_p_13233.html [accessed: 8 October 2011].

Rasmussen, L.L. 2011a. Statsminister Lars Løkke Rasmussens tale ved mindegudstjeneste for Danmarks faldne i Holmens Kirke 5. September 2011. *Statsministeriet* [Online], 5 September 2011. Available at: http://www.stm.dk/_p_13503.html [accessed: 6 September 2011].

Rasmussen, L.L. 2011b. Statsministerens tale i Holmens Kirke. *Venstre.dk* [Online], 5 September 2011. Available at: http://www.venstre.dk/nyheder/enkeltvisning/statsministerens-tale-i-holmens-kirke/ [accessed: 6 September 2011].

Rasmussen, L.L. 2011c. Statsministerens tale ved afsløring af grundstenen i 'Monument for Danmarks Internationale Indsats siden 1948' den 7. April 2011. 7 April 2011. Available at: http://www.stm.dk/_p_13401.html [accessed: 8 April 2011].

Rasmussen, L.L. 2011d. Statsministerens tale ved Folketingets afslutning den 31. maj 2011. *Statsministeriet* [Online], 31 May 2011. Available at: http://www.stm.dk/_p_13463.html [accessed: 1 June 2011].

Rasmussen, M.V. 2008. Flammen og Citronen i Helmand. *Berlingske* [Online], 11 April 2008. Available at: http://www.b.dk/debat/flammen-og-citronen-i-helmand [accessed: 9 May 2011].

Rasmussen, M.V. 2011e. *Den gode krig?: Danmark i Afghanistan 2006–2010.* København: Gyldendal.

Rasmussen, P.N. 2001. Udtalelse af statsminister Poul Nyrup Rasmussen i anledning af terrorangrebet på USA 11 September. Available at: http://www.stm.dk/_p_7165.html.

Rathaus aktuell. 2003. Kabul: Berliner Bär für die Bundeswehr. *Rathaus aktuell* [Online], September 2003. Available at: http://www.berlin.de/rbmskzl/rathausaktuell/archiv/2003/09/03/14806/ [accessed: 11 April 2011].

Ray, J.L. 1995. *Democracy and International Conflict: an Evaluation of the Democratic Peace Proposition.* Columbia, SC: University of South Carolina Press.

Rayment, S. 2007. Army forced to move Ulster memorials. *The Telegraph* [Online], 18 February 2007. Available at: http://www.telegraph.co.uk/news/uknews/1542988/Army-forced-to-move-Ulster-memorials.html [accessed: 13 October 2012].

Rayment, S. 2012. Our big mistake was to assume we had won, says British ambassador to Afghanistan. *The Telegraph* [Online], 10 March 2012. Available at: http://www.telegraph.co.uk/news/worldnews/asia/afghanistan/9135903/Our-big-mistake-was-to-assume-we-had-won-says-British-ambassador-to-Afghanistan.html [accessed: 11 March 2012].

Reichelt, P. 1995. *Politik mit der Erinnerung: Gedächtnisorte im Streit um die nationalsozialistische Vergangenheit.* München: Carl Hanser.

Reinhold, C. 2009. *Danmark i Afghanistan.* København: Forsvarets Mediecenter.

Reiter, D. and Stam, A.C. 2002. *Democracies at War.* Princeton: Princeton University Press.

Rid, T. and Hecker, M. 2009. *War 2.0: Irregular Warfare in the Information Age.* Westport: Praeger.

Riddell, M. 2009. Lord Malloch-Brown interview: 'This dangerous gridlock that taints our politics'. *The Telegraph*, 22 July.

Rigg, C.D. 2010. 15 January 2007. In Mcnab, A. (ed.) *SPOKEN FROM THE FRONT. Real Voices from the battlefield of Afghanistan.* London: Corgi Books 160–71.

Ring, K. 2011. De faldnes monument: SoldaterNyt besøger Finn Reinbothes værk. SoldaterNyt [Online], 4 October 2011. Available at: http://soldaternyt.dk/2011/10/de-faldnes-monument-soldaternyt-besoger-finn-reinbothes-vaerk/#comments [accessed: 9 December 2011].

Ringsmose, J. 2007. Heltene fra Musa Qala. Avisen.dk [Online], 7 June 2007. Available at: http://cms.blogs.ku.dk/2007/06/07/heltene-fra-musa-qala/ [accessed: 5 May 2011].

Ringsmose, J. and Thruelsen, P.D. 2010. NATO's counterinsurgency campaign in Afghanistan: are classical doctrines suitable for alliances? *UNISCI Discussion Papers*, 22(1): 56–77.

Rinke, A. 2010. Die Gnade der totalen Niederlage. Deutsche und Taliban im Vergleich. *Interationale Politik*, 4: 114–21.

Rivenburgh, N. 2000. Social Identity and News Portrayals of Citizens Involved in International Affairs. *Media Psychology*, 2: 303–29.

Rode, R. 2010. Die militärische Niederlage des Westens in Afghanistan. Kluges Niederlagenmanagement und die Chance des Konverzion in einen weltpolitischen Sieg. *Hallenser IB-Papier*, 1.

Roed, F. 2007. Danske soldater jubler over døde talebankrigere. *avisen.dk* [Online], 15 November 2007. Available at: avisen.dk/danske-soldater-jubler-over-doede-talebankrigere_2768.aspx [accessed: 19 January 2011].

Ross, A. 2004. Humanitäre Hegemonie. *Frankfurter Allgemeine Zeitung*, 27 September 2004.

Rottbøl, E. 2011. Konservative vil ændre pointsystem for udlændinge. *Information* [Online], 1 February 2011. Available at: http://www.information.dk/258200 [accessed: 9 May 2011].

Rundschau, F. 2009. Die Gefallenen – eine traurige Liste. *Frankfurter Rundschau* [Online], 15 June 2009. Available at: http://www.fr-online.de/einsatz-in-afghanistan/bundeswehrsoldaten-die-gefallenen---eine-traurige-liste,1477334,2678456.html [accessed: 12 March 2013].

Rundschau, F. 2010. Taliban töten eine Deutsche. *Frankfurter Rundschau* [Online], 7 August. Available at: http://www.fr-online.de/politik/zehn-tote--taliban-toeten-eine-deutsche-,1472596,4538964.html [accessed: 8 June 2011].

Rusch, S.O. 2010. AFGHANISTAN – Ein Land zwischen Krieg und Frieden. *Humanistischer Verband Deutschlands* [Online], 14 January 2010. Available at: http://www.hvd-bb.de/node/462 [accessed: 14 May 2011].

Rutherford, A. 2011. Remembrance: Thousands honour war dead in Northern Ireland. *Belfast Telegraph* [Online]. Available at: http://www.belfasttelegraph.co.uk/news/local-national/northern-ireland/remembrance-thousands-honour-war-dead-in-northern-ireland-16076965.html#ixzz1oFH4VP4u [accessed: 18 November 2011].

Ruus, T. 2009. Kammeraterne tændte lys for Rocco. *Ekstra Bladet* [Online], 9 April 2009. Available at: http://ekstrabladet.dk/nyheder/krigogkatastrofer/article993801.ece [accessed: 11 May 2011].

Rühle, H. 2007. Angst vor der Massenverweigerung. *Süddeutsche Zeitung*, 28 March 2007.

Rytter, N. 2010. Minister: Danmark vinder i Afghanistan. *Politiken.dk* [Online], 13 March 2010. Available at: http://politiken.dk/indland/ECE923195/minister-danmark-vinder-i-afghanistan/ [accessed: 14 March 2010].

Sabrow, M. 2009. Den Zweiten Weltkrieg erinnern. *Aus Politik und Zeitgeschichte*, 36–7.

Schachtman, N. 2010. How the Afghanistan Air War Got Stuck in the Sky. *Wired* [Online], 15 January 2010. Available at: http://www.wired.com/magazine/2009/12/ff_end_air_war/all/1 [accessed: 10 November 2012].

Schake, K. 2012. US retrenchement is right and overdue. In Heisbourg, F., Ischinger, W., Robertson, G., Schake, K. and Valasek, T. (eds) *ALL ALONE? What US retrenchement means for Europe and NATO*. London: Centre for European Reform, 5–26.

Schetter, C. 2007. Die Taliban und die Neuordnung Afghanistans. In Chiari, B. (ed.) *Afghanistan – Wegweiser zur Geschichte*. Paderborn: Militärgeschichtliches Forschungsamt, 77–93.

Schindler, A. 2010. Ich habe einen Freund verloren. In Werner, U.S. (ed.) *'Ich krieg mich nicht mehr under Kontrolle' Kriegsheimkehrer der Bundeswehr*. Köln: Fackelträger.

Schlagintweit, R. 2005. Die deutsch-afghanischen Beziehungen zwischen dem Ende des 2. Weltkrieges und dem 11. September 2001. In Gomm-Ersting, C. and Günther, A. (eds) *Unterwegs in die Zukunft. Afghanistan – drei Jahre nach dem Aufbruch vom Petersberg*. Berlin: Berliner Wissenschaftsverlag, 39–59.

Schmidt, C. 2011. Die Europäisierung der Bundeswehr. *ZFAS, Zeitschrift für Außen – und Sicherheitspolitik*, 4: 173–82.

Schmidt, H.T. 2012. Statsministerens tale ved mindegudstjeneste i Holmens Kirke. 5 September 2012. Available at: http://www.stm.dk/_p_13739.html [accessed: 9 September 2012].

Schmiese, W. 2000. *Fremde Freunde: Deutschland und die USA zwischen Mauerfall und Golfkrieg*. Paderborn: Ferdinand Schöningh.

Schmuda, B. 2009. Trauma, Ehre, Anerkennung? In Hauswedell, C. (ed.) *Soldatentod in heutigen Kriegen. Herausforderungen für politische Normenbildung und Erinnerungskultur*. Rehburg-Loccum: Loccummer Protokolle, 197–9.

Schockenhoff, A. 2012. Einsatz im Afghanistan. Fragen und Antworten. 11 December 2012. Berlin: CDU/CSU-Bundestagsfraktion.

Schreer, B. 2010. Political Constraints: Germany and Counterinsurgency. *Security Challenges*, 6(1): 97–108.

Schröder, G. 2001. Terroranschläge in den USA und Beschlüsse des Sicherheitsrates der Vereinten Nationen. Plenarprotokoll, Sitzung 14/187: 19 September, Berlin: Deutscher Bundestag.

Schuler, K. 2009. Tapfer in Afghanistan. *Die Zeit* [Online], 6 July 2009. Available at: http://www.zeit.de/online/2009/28/bundeswehr-tapferkeit-orden [accessed: 9 October 2011].

Schöllgen, G. 2004. *Der Auftritt: Deutschlands Rückkehr auf die Weltbühne.* Berlin: Ullstein.

Schörnig, N. 2007. Visionen unblutiger Kriege. Hightech-Antworten zur Umgehung der Opfersensibilität. In Geis, A., Müller, H. and Wagner, W. (eds) *Schattenseiten des Demokratischen Friedens – Zur Kritik einer Theorie liberaler Außen – und Sicherheitspolitik.* Frankfurt: Campus, 93–121.

Searle, A. 2007. Last post. *The Guardian* [Online], 12 March 2008. Available at: http://www.guardian.co.uk/politics/2007/mar/12/iraq.art/print [accessed: 19 June 2012].

Sebastian. 2012. BundeswehrVerband wird einen Veteranen-Tag aktiv mitgestalten [Online], 17 February 2012. Available at: http://soldatenglueck.de/2012/02/17/68294/bundeswehrverband-wird-einen-veteranen-tag-aktiv-mitgestalten/ [accessed: 19 March 2012].

Security Council. 2001. Resolution 1386 (2001) On the situation in Afghanistan. New York: United Nations.

Seibel, A. 2013. Die deutsche Zurückhaltung in Mali ist peinlich. *Welt* [Online], 20 January 2013. Available at: http://www.welt.de/debatte/kommentare/article112919350/Die-deutsche-Zurueckhaltung-in-Mali-ist-peinlich.html [accessed: 22 January 2013].

Selchau-Mark, N.C. 2007. Mission mulig: Boykot Ekstra Bladet. *Århus Stiftstidende* [Online], 28 October 2007. Available at: http://stiftenblog.dk/showpost.asp?bid=1361&pid=5002 [accessed: 20 May 2011].

Semler, C. 2007. Jungs gedankenloses Denkmal. *die tageszeitung*, 24 May 2007, 1.

Shakespeare, W. *Much Ado About Nothing.* New York: Barnes & Noble.

Sharp, U.S.G. and Westmoreland, W.C. 1969. *Report on the war in Vietnam, as of 30 June 1968: Section I: Report on air and naval campaigns against North Vietnam and Pacific Command-wide support of the war, June 1964–July 1968.* Washington.

SIPRI. 2012. SIPRI Yearbook 2012 Online. Stockholm: Stockholm International Peace Research Institute.

Skadhede, A. 2007. Helmand er soldaternes mareridt. *DR Nyheder Udland* [Online], 27 September 2007. Available at: http://www.dr.dk/Nyheder/Udland/2007/09/27/062206.htm [accessed: 17 March 2011].

Sky News. 2009. Anti-War Protest Pays Tribute To The Fallen [Online], 17 August 2009. Available at: http://news.sky.com/story/717680/anti-war-protest-pays-tribute-to-the-fallen [accessed: 20 October 2010].

Skytt, L. 2010. Rørende modtagelse af dræbt soldat. *Jyllands-Posten.*

Sledge, M. 2005. *Soldier Dead [How We recover, Identify, Bury, and Honor Our Military Fallen].* New York: Columbia.

Smith, H. 2005. What Costs Will Democracies Bear? A Review of Popular Theories of Casualty Aversion. *Armed Forces & Society*, 31(4): 487–512.

Smith, M., Fenton, B. and La Guardia, A. 2001. American 'Body-Bag Syndrome Is Holding Back NATO'. *The Telegraph* [Online], 21 March 2001. Available at: http://www.telegraph.co.uk/news/worldnews/northamerica/usa/1327353/American-body-bag-syndrome-is-holding-back-Nato.html [accessed: 18 June 2012].

Smith, R. 2006. *The Utility of Force. The Art of War in the Modern World.* London: Penguin.

Smith, T.A. and Tatalovich, R. 2003. *Cultures at War: Moral Conflicts in Western Democracies.* Peterborough, Ont.: Broadview Press.

Snow, L. 2012. We Are Not The Dead: soldiers' faces before, during and after serving in Afghanistan. *The Telegraph* [Online], 31 January 2012. Available at: http://www.telegraph.co.uk/news/picturegalleries/uknews/9013365/We-Are-Not-The-Dead-soldiers-faces-before-during-and-after-serving-in-Afghanistan.html [accessed: 31 January 2012].

Soldatenglück.de. 2010. Verteidigungsminister Guttenbergs und Bundeskanzlerin Merkels Trauer-Reden in Selsingen (Videos). Soldatenglück.de.

Sondhaus, L. 2006. *Strategic Culture and Ways of War.* London: Routledge.

Sophocles. 2003. *Antigone.* Cambridge: Cambridge University Press.

Speicher, S. 2009. Die Namen der Toten. *Süddeutsche Zeitung*, 8 September 2009.

Spiegel. 2001. Interview mit Günter Grass. 'Amerikanische Politik muss Gegenstand der Kritik bleiben'. *Spiegel.online* [Online], 10 October 2001. Available at: http://www.spiegel.de/kultur/gesellschaft/0,1518,161444,00.html [accessed: 16 April 2011].

Spiegel. 2009. Kunduz-Bombardement. Bundeswehr-Generalinspekteur Schneiderhan tritt zurück [Online], 26 November 2009. Available at: http://www.spiegel.de/politik/deutschland/0,1518,663561,00.html [accessed: 18 April 2011].

Spiegel. 2010. Lammert wirft Linke aus dem Bundestag [Online], 26 February 2010. Available at: http://www.spiegel.de/politik/deutschland/afghanistan-lammert-wirft-linke-aus-dem-bundestag-a-680472.html [accessed: 27 February 2010].

Spiegel. 2012. De Maizières Veteranentag sorgt für Wirbel [Online], 16 February 2012. Available at: http://www.spiegel.de/politik/deutschland/bundeswehr-de-maizieres-veteranentag-sorgt-fuer-wirbel-a-815704.html [accessed: 17 February 2012].

Sponsel, R. 2003. Das Embedded Syndrom. Oder: vom Verlust der Objektivität durch embedded journalism. *Internet Publikation für Allgemeine und Integrative Psychotherapie.*

Spranger, M.T. 2000. BVerfGE 93, 266 ff. – Soldaten sind Mörder – Zur Bedeutung der Meinungsfreiheit für die strafgerichtliche Verurteilung wegen der Aussage 'Soldaten sind Mörder'. In Menzel, J. (ed.) *Verfassungsrechtsprechung. Hundert Entscheidungen des Bundesverfassungsgerichts in Retrospektive.* Tübingen: Mohr Siebeck, 592–8.

Steenbuch, J.A. 2008. Fogh: Vi skal bakke op om soldaterne. *Information* [Online], 27 March 2008. Available at: http://www.information.dk/156917 [accessed: 19 May 2012].

Steinmeier, F.-W. 2011. '10 Jahre nach 9/11 – Lehren und Perspektiven' Bekämpfung und Prävention des internationalen Terrorismus. *SPD-Bundestagsfraktion*, 8 September.

Stenstrup, M. and Broberg, M.B. 2006. Forsvarets største opgave. *Jyllands-Posten*, 30 August.

Sterling, K. 2008. Kampsoldater kritiserer Afghanistan indsats. *Information* [Online], 13 July 2008. Available at: http://www.information.dk/162169 [accessed: 19 January 2012].

Stern. 2009. Darum ist die Bundeswehr in Afghanistan [Online], 9 September 2009. Available at: http://www.stern.de/politik/deutschland/einsatz-am-hindukusch-darum-ist-die-bundeswehr-in-afghanistan-1508110.html [accessed: 19 April 2010].

Steuter, E. and Wills, D. 2010. 'The vermin have struck again': Dehumanizing the enemy in post 9/11 media representations. *Media, War & Conflict*, 3(2): 152–67.

Steuter, E. and Wills, D. 2011. Making the Muslim enemy. The social construction of the enemy in the war on terror. In Carlton-Ford, S. and Ender, M.G. (eds) *The Routledge Handbook of War and Society*. London: Routledge, 257–67.

Stirrup, J. 2009. Annual Chief of the Defence Staff Lecture. 3 December 2009. Available at: http://www.rusi.org/events/past/ref:E4B184DB05C4E3/ [accessed: 14 April 2010].

Stoddard, A., Harmer, A. and DiDomenico, V. 2009. Providing aid in insecure environments: 2009 Update. Trends in violence against aid workers and the operational response. *HPG Brief*, 34.

Stougaard, K.E. and Caruso, J. 2009. Soldater jubler over flagdag. *Berlingske* [Online], 31 March 2009. Available at: http://m.b.dk/touch/article.pml?guid=1120239 [accessed: 15 May 2011].

Stougaard, K.E., Caruso, J., Nielsen, L.H. and Nissen, M. 2010. *De faldne*. København: Berlingske Media Forlag.

Stratton, A. 2009. Gordon Brown unveils new plan to target 'crucible of terrorism'. *The Guardian*, 27 April.

Struck, P. 2002. 'Die Sicherheit der Bundesrepublik wird auch am Hindukusch verteidigt.' Verteidigungsminister Struck will die 'Reform der Bundeswehr' fortführen. *AG Friedensforschung* [Online], 5 December 2002. Available at: http://www.ag-friedensforschung.de/themen/Bundeswehr/struck.html [accessed: 14 March 2013].

Struck, P. 2004. Regierungserklärung. *Plenarprotokoll 15/97*. 11 March 2004 ed. Berlin: Deutscher Bundestag.

Struck, P. 2005. Rede des Bundesministers der Verteidigung, Dr. Peter Struck, anlässlich der zentralen Gedenkfeier des Volksbundes Deutsche Kriegsgräberfürsorge zum Volkstrauertag. 13 November 2005. Berlin: Deuscher Bundestag.

Sturm, D.F. 2002. Schröders deutscher Weg. *Die Welt*.

Stuttgarter Nachrichten. 2008. Bundeswehr gerät ins Visier. *Stuttgarter Nachrichten* [Online], 2 August 2008. Available at: http://content.stuttgarter-zeitung.de/stz/page/1804964_0_9223_-nervositaet-am-kundus-bundeswehr-geraet-ins-visier.html [accessed: 19 March 2011].

Svane, A.M. 2007a. Danske soldater bliver i Afghanistan. *Jyllands-Posten* [Online], 17 May 2007. Available at: http://jyllands-posten.dk/indland/article3359713.ece [accessed: 10 June 2012].

Svane, A.M. 2007b. Fogh til soldater: 'Vi gør en forskel'. *Jyllands-Posten* [Online], 18 May 2007. Available at: http://jp.dk/udland/article944398.ece?service=printversion.

Svendsen, J. 2008. Talebans tabstal skal frem i lyset. *Politiken* [Online], 26 October 2008. Available at: http://politiken.dk/politik/ECE588601/talebans-tabstal-skal-frem-i-lyset/ [accessed: 1 June 2010].

Süddeutsche Zeitung. 2008. Die Regierung trägt eine Mitverantwortung [Online], 22 October 2008. Available at: http://www.sueddeutsche.de/politik/bundeswehrverband-die-regierung-traegt-eine-mitverantwortung-1.525288 [accessed: 19 March 2010].

Süddeutsche Zeitung. 2011. Eine Stadt trägt Trauer [Online], 28 February 2011. Available at: http://www.sueddeutsche.de/politik/gefallene-bundeswehr-soldaten-eine-stadt-traegt-trauer-1.1065029 [accessed: 29 February 2011].

Süddeutsche Zeitung. 2012a. Bundeswehrverband zweifelt an Truppenabzug bis 2014 [Online], 1 January 2012. Available at: http://www.sueddeutsche.de/politik/kampfeinsatz-in-afghanistan-bundeswehrverband-zweifelt-an-truppenabzug-bis-1.1247798 [accessed: 4 January 2012].

Süddeutsche Zeitung. 2012b. Linke wirft de Maizière Militarisierung der Gesellschaft vor [Online], 16 February 2012. Available at: http://www.sueddeutsche.de/politik/vorstoss-fuer-einen-veteranentag-linke-wirft-de-maizire-militarisierung-der-gesellschaft-vor-1.1286263 [accessed: 1 March 2012].

Sørensen, I.B. 2010. Tam flagdag gør intet for demokratiet. Videnskab.dk [Online], 3 September 2010. Available at: http://videnskab.dk/kultur-samfund/tam-flagdag-gor-intet-demokratiet [accessed: 19 November 2011].

Tagesspiegel. 2008. Jung entschuldigt sich für tote Zivilisten [Online], 3 September 2008. Available at: http://www.tagesspiegel.de/politik/international/jung-entschuldigt-sich-fuer-tote-zivilisten/1316858.html [accessed: 15 February 2012].

Telegraph. 2001–2012. British war dead in Afghanistan [Online], 24 September 2012. Available at: http://www.telegraph.co.uk/news/newstopics/onthefrontline/5827340/British-war-dead-in-Afghanistan.html [accessed: 13 March 2013].

Telegraph. 2009a. Lieutenant-Colonel Rupert Thorneloe [Online], 3 July 2009. Available at: http://www.telegraph.co.uk/news/obituaries/military-obituaries/army-obituaries/5735309/Lieutenant-Colonel-Rupert-Thorneloe.html [accessed: 17 October 2011].

Telegraph. 2009b. Lt Col Rupert Thorneloe's father says Government fails to put defence first [Online], 31 July 2009. Available at: http://www.telegraph.co.uk/news/worldnews/asia/afghanistan/5951416/Lt-Col-Rupert-Thorneloes-father-says-Government-fails-to-put-defence-first.html [accessed: 8 June 2011].

Telegraph. 2009c. Prince of Wales attends funeral of Lieutenant Colonel Rupert Thorneloe [Online], 16 July 2009. Available at: http://www.telegraph.co.uk/news/uknews/defence/5842625/Prince-of-Wales-and-Duchess-of-Cornwall-attend-funeral-Lieutenant-Colonel-Rupert-Thorneloe.html [accessed: 9 March 2011].

Tenberg, N. 2009. Es ist Krieg – und keiner schaut hin. *Die Tageszeitung* [Online], 2 February 2009. Available at: http://www.taz.de/!29741/ [accessed: 10 June 2010].

Thatcher, M. 1993. *The Downing Street Years*. London: HarperCollins.

The Scotsman. 2008. They put robins on stamps – why not our sons who died in Iraq? [Online], 2 December 2008. Available at: http://www.scotsman.com/news/they-put-robins-on-stamps-why-not-our-sons-who-died-in-iraq-1-1150426 [accessed: 19 December 2010].

The Sun. 2009. Don't you know there's a bloody war on? [Online], 28 August 2009. Available at: http://www.thesun.co.uk/sol/homepage/news/sun_says/2613540/The-Sun-Says.html [accessed: 19 March 2011].

Thrall, A.T. 2000. *War in the Media Age*. Cresskill, N.J.: Hampton Press.

Timmermann-Levanas, A. and Richter, A. 2010. *Die reden – Wir sterben. Wie unsere Soldaten zu Opfern der deutschen Politik werden*. Bonn: Bundeszentrale für politische Bildung.

Toje, A. 2010. *The European Union as a small power: After the post-Cold War*. New York: Palgrave Macmillan.

Townsend, M. 2007. The hardest fight of all for a Falklands hero. *Telegraph*, 24 January.

Townsend, M. 2012. *Point Man*. London: Faber and Faber.

Tran, J. 2010. *The Vietnam War and Theologies of Memory: Time and Eternity in the Far Country*. Chichester: Wiley-Blackwell.

Tran, M. 2009. Profile: Lieutenant Colonel Rupert Thorneloe. *The Guardian* [Online], 3 July 2009. Available at: http://www.guardian.co.uk/uk/2009/jul/03/lieutenant-colonel-rupert-thorneloe [accessed: 17 October 2012].

Travis, A. 2001. 2 in 3 back air strikes. *The Guardian*, 18 September.

Triebel, W. 2010. Afghanistan – letzter Akt der Nato. Die Linke [Online], February 2010. Available at: http://www.die-linke.de/index.php?id=181&tx_ttnews%5Btt_news%5D=11788&tx_ttnews%5BbackPid%5D=154&no_cache=1 [accessed: 14 March 2013].

Tuxen, K.S. 2012. Vi bør lære af Sovjet i Afghanistan [Online], 1 February 2012. Available at: http://www.dr.dk/Nyheder/Udland/2012/02/01/151208.htm [accessed: 3 February 2012].

TV2. 2007a. Gade kritiserer NATO-allierede. 12 October 2007 [accessed: 18 January 2012].

TV2. 2007b. Major Storrud bisættes på onsdag. *TV2 Nyhederne* [Online]. Available at: http://finans-dyn.tv2.dk/nyheder/article.php/id-9019809:major-storrud-bis%C3%A6ttes-p%C3%A5-onsdag.html [accessed: 22 May 2011].

TV2. 2007c. Tyskland klar til tørn i Gereshk. *TV2 Nyhederne* [Online], 30 November 2007. Available at: http://nyhederne-dyn.tv2.dk/article.php/id-9561013:tyskland-klar-til-t%C3%B8rn-i-gereshk.html [accessed: 17 January 2012].

TV2. 2011. Armadillo: 'Rens grøften!'. *TV2 Nyhederne* [Online], 3 January 2011. Available at: http://nyhederne.tv2.dk/article.php/id-35935281:armadillo-rens-gr%C3%B8ften.html [accessed: 5 January 2011].

UD and FD. 2008. *Den danske indsats i Afghanistan 2008–2012*. København: Udenrigsministeriet & Forsvarsministeriet.

UD and FD. 2010. *Den danske indsats i Afghanistan*. København: Udenrigsministeriet og Forsvarsministeriet.

Udenrigsministeriet. 2013. *AFGHANISTANPLANEN 2013–2014. På vej mod fuldt afghansk ansvar*. København: Udenrigsministeriet og Forsvarsministeriet.

Union Européenne. 2003. Une Europe sûre dans un monde meilleur. Stratégie européenne de sécurité. 12 December 2003 ed. Bruxelles: Union Européenne.

Vardrup, J. 2008. Jeg så gerne, at vi var lidt skrappere. *Berlingske* [Online], 27 October 2008. Available at: http://www.b.dk/politik/jeg-saa-gerne-vi-var-lidt-skrappere [accessed: 19 January 2011].

Volker, K. 2011. Don't Call It a Comeback. *Foreign Policy* [Online], 23 August 2011. Available at: http://www.foreignpolicy.com/articles/2011/08/23/dont_call_it_a_comeback?print=yes&hidecomments=yes&page=full [accessed: 24 August 2011].

Værge, J. 2010. Danske soldater slår ihjel – hvor er debatten? [Online], 7 May 2010. Available at: http://www.etik.dk/artikel/365381:Krig-og-etik--Danske-soldater-slaar-ihjel---hvor-er-debatten [accessed: 20 June 2010].

Wagner-Pacific, R. and Schwartz, B. 1991. The Vietnam Veterans Memorial: Commemorating a Difficult Past. *The American Journal of Sociology*, 97(2): 376–420.

Wall, P. 2012. The British Army: Beyond Afghanistan. 31 January 2012. Available at: http://www.iiss.org/recent-key-addresses/the-british-army-beyond-afghanistan/transcript/ [accessed: 5 February 2013].

Wallin, J. 2009. Wootton Bassett rejects talk of Royal title. *Swindon Advertiser* [Online], 16 September 2009. Available at: http://www.swindonadvertiser.co.uk/news/4633151.Wootton_Bassett_rejects_talk_of_Royal_title/?ref=rss [accessed: 19 September 2012].

Watson, J. 2007. Representing Realities: An Overview of News Framing. *Keio Communication Review*, 29(1): 107–31.

Wehrbeauftragter. 2008. Unterrichtung durch den Wehrbeauftragten Jahresbericht 2007 (49. Bericht), Drucksache 16/8200. Berlin: Deutscher Bundestag.

Wehrbeauftragter. 2009. Unterrichtung durch den Wehrbeauftragten Jahresbericht 2008 (50. Bericht), Drucksache 16/12200. Berlin: Deutscher Bundestag.

Wehrbeauftragter. 2010. Unterrichtung durch den Wehrbeauftragten. Jahresbericht 2009 (51. Bericht). Berlin: Deutscher Bundestag.

Weinlein, A. 2007. Gedanken ja – aber wo und wie? Verteidigungsminister Jung weist Kritik an Plänen für Ehrenmal zurück. *Das Parlament*, 9 July, 5.

Welt. 2001. Schröder fordert uneingeschränkte Solidarität [Online], 15 September 2001. Available at: http://www.welt.de/print-welt/article476031/Schroeder_fordert_uneingeschraenkte_Solidaritaet.html [accessed: 19 June 2011].

Welt. 2002. Was verschweigen sie uns noch? [Online], 25 February 2002. Available at: http://www.welt.de/print-welt/article376044/Was-verschweigen-sie-uns-noch.html [accessed: 17 March 2012].

Welt. 2007. Lafontaine warnt vor Terror durch Bundeswehr-Einsatz [Online], 27 July 2007. Available at: http://www.welt.de/politik/article1059243/Lafontaine_warnt_vor_Terror_durch_Bundeswehr_Einsatz.html.

Welt. 2011. Wehrbeauftragter rügt Ausrüstung der Bundeswehr [Online], 3 June 2011. Available at: http://www.welt.de/politik/ausland/article13409492/Wehrbeauftragter-ruegt-Ausruestung-der-Bundeswehr.html [accessed: 17 June 2011].

Wenzel, C. 2010. Robbe will Veteranentag ohne Heldenverklärung. *Die Welt* [Online], 14 July 2010. Available at: http://www.welt.de/politik/deutschland/article8455371/Robbe-will-Veteranentag-ohne-Heldenverklaerung.html.

Wernstedt, R. 2009. *Deutsche Erinnerungskulturen seit 1945 und der Volksbund Deutsche Kriegsgräberfürsorge e.V.* Kassel: Volksbund Deutsche Kriegsgräberfürsorge e.V.

Westdeutsche Zeitung. 2011. Bewegende Trauerfeier für getötete Bundeswehrsoldaten [Online], 3 June 2011. Available at: http://www.wz-newsline.de/home/politik/ausland/bewegende-trauerfeier-fuer-getoetete-bundeswehrsoldaten-1.675567 [accessed: 6 July 2011].

Westerwelle, G. 2010. Rede von Außenminister Guido Westerwelle bei der zentralen Gedenkveranstaltung zum Volkstrauertag, 14.11.2010 in Berlin, 15 November 2010 [accessed: 18 February 2012].

Wevelsiep, C. 2011. Die postheorische Gesellschaft und ihre Freunde. Zum Argument der begrenzten interventionsbereitschaft moderner Gesellschaften. *ZFAS, Zeitschrift für Außen – und Sicherheitspolitik*, 2: 241–62.

White, N. 2009. *Democracy Goes to War. British Military Deployments Under International Law.* Oxford: Oxford University Press.

Whitman, W. 1892. *Prose Works.* Philadelphia: David McKay.

Wieczorek-Zeul, H. 2003. Rede der Bundesministerin für wirtschaftliche Zusammenarbeit, Heidemarie Wieczorek-Zeul, zur Fortsetzung und Erweiterung der Beteiligung bewaffneter deutscher Streitkräfte an dem Einsatz einer Internationalen Sicherheitsunterstützungstruppe in Afghanistan vor dem Deutschen Bundestag am 24 Oktober in Berlin. *Bulletin der Bundesregierung*, 91–3.

Wieczorek-Zeul, H. 2008. Frieden durch Entwicklung. *Argumente*, 1: 38–41.

Wilson, J. 2009–2010. Recovering the Dead. *British Army Review*, 148: 106–7.

Winter, J. 1995. *Sites of Memory, Sites of Mourning: The Great War in European cultural history.* Cambridge: Cambridge University Press.

Winter, J. 2006. *Remembering War.* New Haven: Yale University Press.

Worm, N. 2007. Gade: Major tog selv beslutningen. *Jyllands-Posten* [Online], 17 October 2007. Available at: http://jyllands-posten.dk/indland/article3292355.ece [accessed: 5 October 2011].

Wulff, C. 2010. Rede, Zentrale Gedenkfeier zum Volkstrauertag im Deutschen Bundestag. 14 November 2010. Berlin: Deutscher Bundestag.

Wulff, C. 2011. Feierlicher Appell zur Ablegung des Gelöbnisses von Soldatinnen und Soldaten der Bundeswehr. *Der Bundespräsident* [Online], 20 July 2011. Available at: http://www.bundespraesident.de/SharedDocs/Reden/DE/Christian-Wulff/Reden/2011/07/110720-Bundeswehrgeloebnis.html [accessed: 21 July 2011].

Zaller, J.R. 1992. *The Nature and Origins of Mass Opinion.* Cambridge: Cambridge University Press.

Zeit. 2007. Ehrenmale. Der Dank des Vaterlandes [Online], 30 May 2007. Available at: http://www.zeit.de/online/2007/22/bg-ehrenmale#10 [accessed: 19 June 2012].

Zelizer, B. 2010. *About to Die. How News Images Move the Public.* Oxford: Oxford University Press.

Zøllner, K. 2008. *Du må ikke slå ihjel.* København: Lindhardt og Ringhof.

Aallmann, T.Ø. 2011. *Løvehjerter. Med panserinfanteristerne i Helmand.* Odense: Trykkefrihedsselskabets Bibliotek.

Index

For Product Safety Concerns and Information please contact our EU
representative GPSR@taylorandfrancis.com
Taylor & Francis Verlag GmbH, Kaufingerstraße 24, 80331 München, Germany

www.ingramcontent.com/pod-product-compliance
Ingram Content Group UK Ltd.
Pitfield, Milton Keynes, MK11 3LW, UK
UKHW011014180926
475701UK00004B/741

* 9 7 8 1 1 3 8 2 7 9 3 5 3 *